AF601568

Introductory Ornamental Horticulture and Landscape Gardening

The Authors

Dr Rajaneesh Singh is presently working as Associate Professor and Head, Department of Horticulture, Tilak Dhari Post Graduate College, Jaunpur (U.P.). He did his Graduation in Agriculture from Udai Pratap Autonomous College, Varanasi, Post Graduation in Horticulture from G.B. Pant University of Agriculture and Technology, Pantnagar and Ph.D. in Horticulture from BHU, Varanasi.

Dr Singh has served Tilak Dhari Post Graduate College in various capacities like In-charge Garden Section, Captain in NCC, Member of the Sports Committee, Academic Counselor of IGNOU study center and Former Proctor of the college. Dr. Singh has 19 years teaching experience of horticulture. Dr. Singh has more than 62 publications to his credit (research papers 25, proceeding papers 5, conference/seminar papers 10, books 2, book chapters 10 and popular/semi-scientific articles 10). He had been chairman of the organizing committee of One National Workshop. Dr. Singh qualified ICAR NET in Horticulture and Vegetable Science and received Senior Research Fellowship in ICFRE during Ph.D. programme. He is fellow of Indian Society of Vegetable Science and Indian Society of Horticulture Science. He has supervised 4 M.Sc. (Ag.) and 3 Ph.D. students as Chairman of Advisory Committee.

Dr Bijendra Kumar Singh is presently working as Assistant Professor, Department of Horticulture, Tilak Dhari Post Graduate College, Jaunpur (U.P.). He did his Graduation in Agriculture from VBSPU, Jaunpur, Post Graduation in Horticulture from BBAU, Lucknow and Ph.D. in Horticulture from BHU, Varanasi. He has specialization in Pomology.

Dr Singh has served Tilak Dhari Post Graduate College in various capacities like Member of Campus Greenery of college under IQAC committee and Academic Counselor of IGNOU study center. Dr. Singh has 143 publications (research papers 30, proceeding papers 3, conference/seminar papers 50, book chapters 25 and popular/semi-scientific articles 35). He had been convener of the organizing committee of One National Workshop. Dr. Singh qualified ICAR NET in Fruit Science and received University Grant Commission (UGC) fellowship during his Ph.D. He is a life member of renowned societies and journals in India like Asian PGPR Society of Sustainable Agriculture, International College of Nutrition's, Society for Advancement of Research on Pomegranate, Trends in Biotechnology and Biological Sciences, Advances in Life Sciences and Trends in Biosciences. He has supervised 4 M.Sc. (Ag.) and 3 Ph.D. students as Chairman and Member of Advisory Committee.

Introductory Ornamental Horticulture and Landscape Gardening

Rajaneesh Singh

Bijendra Kumar Singh

2020

Daya Publishing House®

A Division of

Astral International Pvt. Ltd.

New Delhi – 110 002

ISBN 9789390371952 (Int. Edition)

Publisher's Note:

Every possible effort has been made to ensure that the information contained in this book is accurate at the time of going to press, and the publisher and author cannot accept responsibility for any errors or omissions, however caused. No responsibility for loss or damage occasioned to any person acting, or refraining from action, as a result of the material in this publication can be accepted by the editor, the publisher or the author. The Publisher is not associated with any product or vendor mentioned in the book. The contents of this work are intended to further general scientific research, understanding and discussion only. Readers should consult with a specialist where appropriate.

Every effort has been made to trace the owners of copyright material used in this book, if any. The author and the publisher will be grateful for any omission brought to their notice for acknowledgment in the future editions of the book.

Published by : **Daya Publishing House®**
A Division of
Astral International Pvt. Ltd.
– ISO 9001:2008 Certified Company –
4736/23, Ansari Road, Darya Ganj
New Delhi-110 002
Ph. 011-43549197, 23278134
E-mail: info@astralint.com
Website: www.astralint.com

Digitally Printed at : **Replika Press Pvt. Ltd.**

World Noni Research Foundation

64, Third Cross Street, Second Main Road, Gandhi Nagar, Adyar, Chennai - 600 020
Telefax : 044-2442 3601 E-mail : mail@worldnoni.org Website : www.worldnoni.org

Dr. KIRTI SINGH
FNASc., FNAAS, FNABS
CHAIRPERSON

FOREWORD

The association of flowering plants and human society is known since the ancient time. Human being is always attracted by beautiful and fragrant flowers and plants. The floriculture and landscape gardening have emerged as a full fledged discipline all over the country for students of horticulture in various colleges and universities. It has immense potential is increasing income and standard of living of flower growers and traders. The present book entitled "**Introductory Ornamental Horticulture and Landscape Gardening**" is helpful in teaching and learning of undergraduate and postgraduate students of horticulture. This book covers all the introductory aspect of science of ornamental and landscape gardening. In recent year poly-house cultivation of flower is becoming popular in India and use of flowers for various purposes is becoming very popular.

I congratulate and appreciate the effort made by authors to put together the information on various aspects of the subject. I am sure that the publication will be very useful to the students, teachers and even farmers associated with ornamental horticulture and landscape gardening.

Date- 05-09-2019
Place- Jaunpur

(Kirti Singh)
Former Vice-Chancellor

Preface

India is blessed to have a long history of floriculture and gardening since time immemorial. Now-a-days floriculture is regarded as a viable diversification from the traditional field crop due to its higher return per unit area. The floriculture industry comprises the florist trade, nursery, flowering and ornament of plants, bulb and seed production, micro-propagated materials, *etc*.

The present book Introductory Ornamental Horticulture and Landscape Gardening cover all the basic chapters related to ornamental horticulture and landscape gardening. This book covers the syllabus floriculture and landscaping of B.Sc. (Ag.) and M.Sc. (Ag.) horticulture course.

We fill immense pleasure to express our heartfelt gratitude to Prof. Kirti Singh Former Vice-Chancellor for his inspiring guidance encouragement and blessing.

We have immense pleasure to express our gratitude to management, principal and faculty member of agriculture and horticulture for their moral support and encouragement.

We wish to express our appreciation to Dr. Bijendra Kumar Singh (Asst. Professor, Horticulture) for his dedication in compilation and some valuable suggestion regarding the manuscript.

We don’t have to express our emotional feeling towards our parents, family members and relative for their encouragement and support who have been affectionate, understanding and patient during the long hours we spent on the manuscript.

Dr Rajaneesh Singh

Dr Bijendra Kumar Singh

Contents

Section II
Commercial Flower Production

Section-I

Basics of Ornamental Horticulture and Landscaping

Chapter 1

Introduction

The term **Horticulture** is derived from the **Latin** word where ***Hortus*** means ***Garden*** and ***Cultura*** means ***Cultivation***. Which is the also study of fruits, vegetables, flowers and medicinal and aromatic plants. In other word horticulture is the study and cultivation of fruits, vegetables, flowers, medicinal and aromatic plants for commercial or aesthetic purpose.

Father of Horticulture= Thomas Andrew Knight, John Lindley and Liberty Hyde Baily

Father of American Horticulture= Liberty Hyde Baily

Father of Indian Horticulture= Dr. M. H. Marigowda

Father of Systematic Pomology= Decandole

Father of Olericulture= Liberty Hyde Baily

Father of Floriculture= G.S. Randhawa and B.P. Pal

Father of Orchidology= John Lindley

Branches of Horticulture

1. **Pomology-** The term **Pomology** is derived from **Greek** word *Pomum* means **Fruit (Latin word)** and *logy* means to **Study (Greek word)**. It is the study of fruit plants.
2. **Olericulture-** The term **Olericulture** is derived from the **Latin** word *Oleris* means **Pot Herb or Herbaceous** and *Cultura* means **Cultivation or Growing** or **Raising.** It is the study and cultivation of vegetable crops.

3. **Floriculture-** The term **Floriculture** is derived from **Latin** word *Florus* means **Flower** and *Cultura* means **Cultivation**. It is the study of flower crops, while the study of flower is called anthology. The term **Anthology** is derived from the **Greek** word *Anthos* means **Flower** and *Logy* means **Study**.
4. **Ornamental-** These plants are grown for decorative purpose in garden and landscape.
5. **Medicinal plant-** These are used for yielding alkaloid and steroids principle which have preventive and curative properties, *e.g.*, Ashwagandha, Sarpgandha, *etc.*
6. **Aromatic plant-** These plants yield aromatic essential oil on steam distillation or solvent extraction, *e.g.*, Geranium and Jasmine, *etc.*
7. **Spice-** These plants and their products as food adjuncts to add aroma and flavor to food, *e.g.*, Pepper, Cardamom, Clove, Cinnamon, *etc.*
8. **Condiments-** These are used to add only taste to the food, *e.g.*, Coriander, Cumin, Turmeric, Ginger, *etc.*
9. **Plantation crops-** The crop like coconut, areacanut and tea are grown in estate or large acreage or area are called plantation crops. The product of these plants can be used only after processing.
10. **Post-harvest technology-** Horticulture products are transported for longer distance then used. The fermented storage method is suitable for longer uses.

Branches of Ornamental Horticulture

1. **Floriculture-** Growing of flower plants for aesthetic on economical purpose.
2. **Arboriculture-** Growing of tree for aesthetic purpose.
3. **Turf grass science-** The science and art of developing turf grass for lawn purpose.
4. **Landscaping-** The growing of ornamental plants in association with other aesthetic feature or elements to beauty a demarcated area of land.
5. **Interior scaping-** The science and art of indoor plants growing to beautify the interior of a building.

Some Important Terminology on Floriculture

Cut flowers- Cut flowers are widely used in floristry for decoration purpose and to express human sentiments with the increasing aesthetic value of flower in the mind and hearts of people, the demand of cut flowers in the domestic as well as export market has increased many *e.g.* Rose, Annual Chrysanthemum, Gladiolus, Tuberose, Gerbera, Orchids, Carnation, *etc.*

Loose flowers- The loose flower or stem less flowers are widely used for making garland, veni, decoration or rangoli, *etc.* These are the traditional flowers grown in India like Marigold, Jasmines, Gomphrena, *etc.*

Floriculture- Floriculture or flower farming is a discipline of horticulture concern with cultivation of flowering and ornamental plant for garden and floristry comprising the floral industry. The development of plant breeding, the new variety is a major occupation of floriculturist.

Floriculture is the branch of horticulture which deals with commercial growing, marketing and arranging flowers and ornamental plants which includes annuals, biennials and perennials, *viz* tree, shrubs, climber and herbaceous perennial plants. In other word floriculture is growing of cut flowers, potted flower and foliage plant, bedding plant in greenhouse and in field condition. There are several thousand species of flower or plant that grow as commercial crop.

Floriculturist- A person involved in growing, improving and teaching of ornamental horticulture is known as floriculturist.

Florist- A person dealing the business of cut flowers, cut greens, loose flowers and floral product is known as florist.

Deciduous plant- Plant drops its leafs annually.

De-foliage- Removal of foliage or leaf is known as the defoliation for inducing the flowering in plants.

Dense shade- It is also called deep shade and heavy shade, no direct sun light all day with very little reflected indirect light, *e.g.,* Shade under large, fully mature evergreen, shade under raised decks and shade under large free with dense canopies such as live oak.

De-shooting- When the sides shoots after pinching are 3-5cm long then retain 3-5 shoots per plant in standard cultivars. This process is also found in carnation flower plant.

Dis-budding- It is the removal of visible (5-10mm diameter) undesirable buds. This process is also found in carnation flower plant.

Dry flowers- Dry flower arrangements are becoming very popular now a day as a home décorowing to its non-perishability and longer life than fresh flowers like Alliums, Celosia and Zinnia, *etc.*

Essential oils, floral extracts and other products- Essential oils and perfumery from natural sources are in great demand worldwide like rose water and attar.

Evergreen plant- Plant in which foliage is always present.

F_1 hybrid seed production- New and much attractive hybrid varieties in several flower crops can be developed through production F_1 hybrid seeds that have characteristics like dwarfness, compactness, free blooming earliness and long flowering duration like Primula, Petunia, Cyclamen, Marigold and Begonia, *etc.*

Floral carpet- Art depicting a carpet design in the ground with the help of floral materials.

Flower seed- Seasonal or annual flowers are much in demand all over the world for beautifying landscapes. In past, most of the flower seeds were imported from other countries. Subsequently, several leading flower seed production companies set up production facilities in India for targeting domestic market as well as exports.

Foliage plant- Any plant grown primarily for its foliage and utilized for internal decoration or interior landscape purpose. It may have flowers but these may be secondary as compare to foliage feature., *e.g.,* Flowering maple, aglaonema, alternanthera, alocasia, cast iron plant, elephant ear, ferns (boston), philodendron, spider plant, *etc.*

Garden- Garden is a place for growing of horticultural plant (fruit, vegetable, flower and others, *etc.*).

Hanging basket- Training of plant in containers which is suitable for indoor and outdoor conditions, *e.g.,* Petunia and zebrine, *etc.*

Hardening- It is a treatment given immediately after the harvesting of flowers by using water (preferably warm do-ionized water containing some germicide) to restore turgidity.

Hi-tech floriculture- The technology which is ultra modern, less environment dependent and capital intensive having improved productivity with quality produce is high-tech floriculture.

Importance of floriculture- The floriculture in India comprises the florist trade, production of nursery plant, potted plant, bulb and seed production micro propagation and extraction of essential oil from flowers.

Indoor gardening- Indoor gardening is the art of growing and arranging plants indoor or in the house for its best use for function and or beauty.

Landscape design- It is the determination of character of an object to save certain purpose known in advance. The landscape design is aesthetic (attractive in appearance) or economical (utility purpose).

Landscape garden- It is design with definite use of plant to serve certain aesthetic or utilitarian purpose. It is a man-made creation for ornamental or practical or both the uses.

Landscape gardening- It is a planning and planting of outdoor space to secure the most desirable relationship between the land for architect and plan to best meet the human needs for beauty and functional uses.

Landscape- It is general outlaying or appearance of an area in relation to its surrounding.

Landscaping- The design and alternation of a portion of land by use of planting materials and land reconstruction is known as landscaping.

Light shade- 2-3 hr without direct sun as also called thin filtered shade, 10 am to 6 am in summer when sun is most intense there is either

Moisture loving plant- Snake plant (*Sansevieria* sps.), ZZ plant (*Zamioculcas zamiifolia*), Jade plant and Aloe plant are examples of these loving plants, many of these plants are considered succulents a large group of low moisture ornamental plants. The plant native to extremely dry area including cactus such as cereus don't have any leaves at all instead they have a modified stems that hold abundant moisture and carry out photosynthesis. The amount of soil moisture a plant requires for optimal health varies from plant to plant. Similarly to light and the forest canopy concept various type have evolved and adapted to different environmental conditions based on moisture availability. Indoor plant that require less soil moisture have developed modified plant parts and structure the help them cope with drier conditions similar to the way camels have evolved and adopted to dry conditions many plant are thick waxy leaves with flower stomata effective at strong water and reducing water loss.

Natural dyes- Flower are also an important source of natural colours that are largely used for making dyes as an alternative of artificial dye colours. For natural colour dahlia, daylily, hibiscus, marigold, *etc.*

New potential flower crop- There are several newly introduced flower crops that can be used as an alternative of existing flower crops due to their higher genetic diversity and better establishment like Kangaroo Paw, Canna, Lily, *etc.*

Nursery industry- There is huge demand of quality planting material in the market. Using potted plant production, landscape and home gardening as well as commercial production.

Ornamental horticulture- Ornamental horticulture is the branch of horticulture which includes both the aspect of floriculture and landscaping.

Partial shade- 4-5 hr partial sun without direct, as also called dappled shade half shade, medium shade, semi shade, 10 am to 6 am in summer when sun is most intense there is either

Full shade area- Less than 2 hr of direct sun a day, shade all day, reflected direct light

Partial sunny area- 2-5 hr of direct sun or full day of dappled sun a day

Pinching- Removal of shoots apex to overcome apical dormant and promote lateral shoot development.

Plug plant production- Plug plants are young plants raised in small individual cells, ready to be transplanted in the field or containers or garden. Flowers like gerbera and marigold, *etc.* grown as plug plant performs very well after establishment in the field.

Pot plant- Green or foliage plants as well as bedding annual plants having attractive flowers colour and leaves.

Pot poris- Pot poris a special dried floral arrangement is the mixture of sweet smelling leaves, spices, seeds, roots and distilled essential oil which is filled in pillows or sachets.

Pulsing or loading- It consist of placing the lower portion of cut flower stems in solution containing high percentage of sugar and germicide for a period of few hours or two days specific formulation developed vary with the flower species as sucrose 2-20 per cent (2-5 per cent for rose, chrysanthemum and 15-20 per cent for gladiolus, gerbera) for 12-48 hour at 20-27°C and RH 80-100 per cent under 2000-2500 Lux cool light.

Senescence- Ageing of plant parts such as the flower usually the stage from full maturity to death. Senescence is the final phase in ontogeny of the organ in which a series of normally irreversible events initiated that leads to cellular break down and death of organ.

Shade loving plant- Think wood land floor use liberal amount of peat, compost, organic matter, when planting these plants avoid over watering and provide good air circulation for healthy plant. The plants on this list are for full (less than 3 hr sun) filtered and partial shade (3-6 hr direct sun a day) filtered is dappled sunlight through the tree (this is the easiest shade to grow in) avoid planting and these plants in full sun to protect sun scalding damage.

Terrace gardening- A terrace is a raised space of ground constructed around a dwelling house or at the corner of a garden or on the side of a hills. When this terrace is used for some sort of gardening this is known as terrace gardening.

Tissue culture plant production- The demand for flower increasing day by day globally. Plant tissue culture activities are confined to production of ornamental and flowering plants having large global market.

Turf grass- Turf grasses beautify parks and landscapes. They are part of a larger green industry that improves the physical and mental health of the people, with the increases in interest for maintaining lawns, turf grasses are being produced by growers and have because a new avenue of earning profit through selling of turf grasses.

Vase life- The period for which flower or foliage remains is presentable form without losing its grade and quality is known as longevity, vase life, display life or self life. However self life is mostly used in case of cut flower.

Wintering- The base of the plant is exposed to sun and air by removing 10-15cm deep soil layer in 20-30cm plant periphery. It process is also in the rose flower plants.

Chapter 2

Importance and Scope of Floriculture and Landscape Gardening

Introduction

Flowers are associated with mankind from the dawn of cultivation. Flowers are used for various purposes in our day to day life like worshipping, religious purpose, social function, wedding, interior decoration, indoor gardening and self purpose. Flowers commonly used for such purpose are- Rose for love, White carnation for woman love, French marigold for jealousy, African marigold for vulgar minds, Jasmine for amiability, Lily for purity, Pansy for thoughts and Sweet pea for departure, *etc.*

The role of flowers and gardens in the life of the nation and the people is not properly understood in India. Largely, floriculture and gardening are being neglected by the administrators and get last priority in any process of planning or allotment of funds. Even whatever meager funds are allocated to floriculture it is only this branch that is made the first target of any economy drive. It is also sad to note that though we have at least one or more Agricultural Universities in each State, none of them have a professor of floriculture and landscape gardening. Though the Indian Council of Agricultural Research has done a lot for floriculture, one of its leading institutes, the Indian Agricultural Research Institute, New Delhi, which has bred several beautiful rose cultivars, does not have a full-fledged Division of Floriculture and Landscape Gardening. But such a division has been established at the Indian Institute of Horticultural Research, Bangalore, where a number of cultivars of hibiscus, croton, and bougainvilleas have been released for cultivation.

Though floriculture and gardening have to go a long way to develop to a significant level, fortunately the position is not as bad as it used to be some time back. The credit for giving floriculture and garden a good start and consequently some status goes largely to the noted administrator-cum-scientist, Dr. M.S. Randhawa. His work in this field was carried further by another noted scientist and great lover of flowers, particularly roses, Dr. B.P. Pal. But for the encouragement given by these two scientists, floriculture would not have reached the present level of achievement and gained the status that it is having today. Of course, the patronization and encouragement given to floriculture and gardening by the late Prime Minister Jawaharlal Nehru and also the late President Dr. Zakir Husain must also be recalled with gratitude.

Floriculture

It is not possible to define floriculture in the sense as one can define a triangle in the language of geometry. But possibly the closest definition is "Floriculture is the art and knowledge of growing flowers to perfection". But this definition is also not complete, floriculture includes not only flowers but also many ornamental foliage plants, cacti and others. The art of growing bonsai, the dwarfing of plants, also come under floriculture. Often the term 'Ornamental Horticulture' is used to include the culture of flowers, foliage plants, trees, and possibly gardening also. But if the term 'Floriculture' is preferred it should mean not only the culture of flowers but also other ornamental plants such as foliage plants, trees, shrubs, palms, ferns, bamboos, cacti, other succulents, landscape gardening, *etc.*

Significance and Importance

Though many people are nostalgic about floriculture, there are others with brush away the idea as a luxury or even wastage of money and of floriculture cannot be ignored or bypassed so lightly. Flowers symbolize purity, beauty, peace, love, and passion. To a Japanese flower-arranger each flower expresses one or more meanings. To an Indian, especially to Hindus, flowers have a much greater significance. A devoted Hindu needs flowers every morning for religious offering to the family deity. The aesthetic value of flowers in our daily life cannot be over-emphasized. Floral garlands, gajras and venis are needed for marriage ceremonies. The use of gajara and veni is not limited to such occasions as marriage only; these are also used as adornment for hair by our women of all ages, especially in the South, every morning, evening, or even all through the day. Floral ornaments, bouquets, or flower arrangements also find a pride of place in social gatherings, birthday parties, welcoming a home-coming friend or relative and honouring dignitaries. The arrival of newborns is rejoiced with flowers, the sick are wished speedy recovery by offering flowers, while the dead are hidden farewell with flowers along with tears of sorrow. Even those who dismiss floriculture as a luxury cannot afford to do without flowers when the occasion comes.

The potential of floriculture as an industry has not been exploited properly. Floriculture is an intensive type of agriculture and the income per unit area from floriculture is much higher than any other branch of agriculture. In markets such as Delhi and Bombay a single spike of gladiolus may sell for up to five to ten rupees. The business is said to be quite profitable. The nationalized banks provide advance loans to the cultivators to promote this cut flower industry. Rose is another important commercial flower the cut blooms of which are highly prized in the large cities. A dozen good rose stems may sell for up to ₹36 in a city like Bombay. *Rosa damascena* and Edouard rose are grown in Kannauj, Jaunpur, and Ghazipur in U.P. for the manufacture of attar and rose water. The sale of loose flowers of jasmines is a roaring business, especially in Southern India and also in Bombay and Calcutta. The flowers are used mostly for making gajra, veni and garland, as well as for worship.

Role of Floriculture in a Developing Country

In India, floriculture is not getting the priority it deserves, though it has a great role to play, as shall be clear from the foregoing discussion. It is an intensive type of agriculture and the income per acre is much higher than any other agricultural product if it is done in a scientific way. First, we should not neglect the great aesthetic value of flowers in our daily life. Growing colourful flowers in the house compound and in the parks will greatly enhance the beauty of the surroundings in the countryside, the towns, and the cities. They will also meet the daily necessities of our life such as adorning the hair or for indoor decoration or for offering to gods and goddesses. Flowers also bring happiness to life and boost the sagging spirits. Our old cities and towns are shabby-looking. If the citizens and the municipalities encourage the growing of flowers, at least partly, the cities will look much gayer. This is truer for the new industrial townships that are coming up so frequently and which offer much greater opportunity for growing flowers for beautification and uses.

Commercially, floriculture can open up great opportunities to our poor farmers. Our country has diverse climatic conditions which offer the scope for growing several kinds of commercial flowers. The cultivators can deploy a part of their land for growing commercial and common flowers such as marigold, China aster, *etc.*, which do not require much care and generally earn more profit than many other crops. The government should recognize the flower growers into societies and offer them help in selling their produce through a network of retail stores in the big city hotels and markets, thus eliminating the menace of middlemen. Recently, quite a good number of plants are being exported. One vital step in this direction will be to set up an organization for certifying the seeds and standardizing the nursery stocks. At present there is no such organization and the customers are cheated with sub-standard products. Therefore, efforts should be made to disseminate modern methods of cultivation to increase the yield. Efforts should also be made to standardize improved crop production technology in commercial flowers to obtain maximum yield compatible with high quality blooms.

Scope for Floriculture

In India, floriculture is only a developing subject and as such offers much scope for improvement. This problem can be tackled from several angles.

Conservation, Domestication and Introduction

India is rich in its plant resources, many of which are of ornamental value and some are potentially ornamental. Much of this wealth is wasted as a result of rapid urbanization, industrialization, and unscrupulous collection. For example, some unscrupulous nurserymen are selling one beautiful *Paphiopedilum insigne* orchid plant for the high price of a single flower of this species in foreign markets. As a result of such wanton collection many species of orchids are threatened with extinction. To conserve such rare orchids, the I.C.A.R. has now started some orchid sanctuaries, which is a correct step in this direction.

Domestication of wild plants with potential ornamental value is another way of improving garden wealth. In the process of domestication, possibly in an altogether different climate, the wild plants generally pass through many changes, which itself may cause some chance improvement. We have some very good plants in the wild with potential ornamental value, especially ferns, which can be acclimatized and domesticated under quite different climates. In this regard, the National Botanical Research Institute (formerly the National Botanic Gardens), Lucknow, has done some good work in acclimatizing the Himalayan ferns in the sub-tropical climate of Lucknow.

Prospects for Improvement

(a) Hybridization

Acclimatization and introduction are not enough for improving the plant wealth. The tastes of the people change very frequently, and there is always the eraze for new things. One of the best ways to improve is through hybridization between varities of species, or interspecific hybridization within a genus or even intergeneric hybridization. For example, in orchid hybrids in between the gerera Cattleya, Brassarola, and Laelia are available. The hybridization work done in India, especially on roses, has enriched our floriculture to a great extent. Similarly, new hibiscus cultivars released by the Indian Institute of Horticultural Research, Bangalore, and others also enriched our collection of ornamental plants. Our efforts in this regard should be directed towards developing new hybrids suiting the different agro-climatic conditions prevailing in the country.

(b) Mutation

Another important way of developing ornamental wealth is by mutation breeding. The natural mutant bougainvillea 'Mary Palmer' evolved spontaneously in a Calcutta garden has been acclaimed all over the world. But since natural mutants develop only by chance, other physical and chemical mutagens could also be employed to get new attractive mutants. Again, mutation can also be combined with hybridization to further enrich the floricultural wealth.

(c) Polyploidy

Polyploidy is another method of plant improvement. Polyploidy can be induced by colchicine and other chemicals. Polyploids generally have larger flowers and intense-cloured petals sometimes with rolling at the edges. Polyploidy combined with hybridization may result in remarkable improvement.

(d) Propagation

Easy and rapid propagation of ornamentals will go a long way in spreading the cult of floriculture in India. The use of mist propagation units along with the application of root-promoting hormones has enabled many hitherto difficult to root tree and shrubs to root from cuttings. This method has improved the scope of supplying rare plants to garden lovers at comparatively cheap rates. Another field where not much work has been done is the production of disease-free plants. The tissue culture method offers the production of virus-free plant material. This method should be employed increasingly in ornamental plants not only for getting disease-free stocks but also to have rapid multiplication. For example, the process of multiplying orchids and *Anthuriurms* through vegetative means is very slow and once a mother plant is divided for propagation it takes some time to flower again. Moreover, only a limited number of plants can be raised by this method. If the technique of tissue culture is employed, rapid multiplication is possible.

(e) Dwarfing of Plants

Dwarfing of ornamentals by growth-retardant sprays is being exploited commercially in advanced countries. In India also growth regulators have been used to dwarf ornamentals. In the present-day society more and more people living in flats in cities where the scope of gardening is limited or is not there at all. With the help of growth retardants it is the flat dwellers in cities can grow them in their limited spaces. Plants sprayed with growth retardants become tolerant to adverse growing conditions, the leaves become darker in shade and shining, and on many occasions early flowering is induced. In *Chrysanthemum*, the growth-retardant spray can obviate the process of pinching.The commonly used growth retardants are B-Nine, Phosphone, and Cycocel. Maleic hydrazide (MH) in high concentration: (10,000-15,000 ppm) also acts as an effective retardant. This method should be increasingly employed by the nurseries to encourage flat dwellers to take to gardening.

(f) Extending Blooming Period

It often happens that some of the commercial flowers bloom for only a short period of the year irrespective of the planting date. For example, in the North, chrysanthemum starts blooming by November 15 and completes flowering by the first week of January. This causes a glut in the flower market. Therefore, staggering of the blooming period of such crops will go a long way in solving the marketing problem for the growers. The blooming of chrysanthemum can be prolonged by manipulating the photoperiod. This crop needs longer photoperiods for vegetative growth but shorter photoperiods for flowering. It is possible to extend the blooming period of chrysanthemum from the usual 45 days to five months under Lucknow conditions.

Scope and Prospect for Domestic and Export Market

(a) Domestic Market

A conservative estimate puts the land under floriculture in India at around 3,000 hectares. Considering the size and population of our country, this area is not at all impressive. There is scope for putting more area under flowers provided modern scientific methods of cultivation and marketing are adopted. If the volume of trade of nursery stocks such as potted plants, trees, shrubs, bulbs, and seeds are also taken into consideration, the total turnover in the trade of ornamnentals will be quite significant.

In the South, the small-flowered chrysanthemum, especially popular and the turnover is quite large. Research may be directed towards producing a high-yielding strain acceptable to the producer and the consumer. At the Indian Institute of Horticultural Research, Bangalore, some promising small-flowered hybrids have been developed, which may be released shortly for commercial cultivation. The cultivators should also be helped by such research institutes to prolong blooming periods to avoid market gluts. Similarly, in other popular cut flowers such as tuberose, jasmine, china aster, *etc.*, research should be done in this direction.

A proper method of packaging and prolonging the cut flower life will also help enlarge the domestic and the foreign markets. There is unlimited scope for increasing the sale of nursery stocks such as budded roses, indoor plants, bulbs, and seeds, provided the customers are assured of good quality materials. The immediate task will be to set up a certification organization to standardize the quality of these products.

Some of the ornamental plants yield products which are eaten as vegetable or otherwise. There is ample scope to popularize these products and consequently increase their cultivation. For example, the underground stem of *Nelumbium* (lotus), commonly called bhasinda, is a popular vegetable in North India. The same product is also used to make pickles. The stems of water lilies are also eaten as vegetable by many Bengalees, especially those from Bangladesh. The prickly aquatic plant resembling lotus, are commonly found to grow in the tanks of Bangladesh and parts of India, yields seed of the size of a pea which when parched or roasted over hot sand gives a puffed-up product, commonly called makhana, and is a delicacy for the table. The fresh seed of lotus fruit is popularly eaten in Bengal after removing the green plantlet from inside which is bitter, but the seed itself is tasty. With so many water tanks in the Eastern part of the country and elsewhere it is possible to extend the cultivation of these products commercially.

(b) Export Market

The possibility of exporting cut flowers, plants, bulbs, *etc.*, has not been explored properly though there is a very good potential. The Netherlands, and Italy, imported cut flowers worth over ₹88 crores of which the share of West Germany alone was over ₹70 crores. Unfortunately, among the exporting countries the name of India was absent, although we have got favourable and diverse climatic conditions ideal for growing varied types of flowers required by the westerners. The European when these countries remain largely under snow and hence field production of flowers is not possible. In fact, during the winter flowers are produced under glass which increases the total production cost, adding to the already high cost of labour there. In most parts of India the winter is the best period for producing most of the commercial flowers. Therefore, with our abundance of cheap labour we are in an enviable position to build our cut flower trade with these countries. The export of ornamental bulbs, when they are dormant, offers good scope as these can be transported by sea instead of by air and thus can be sold cheaper than fresh flower which are to be air-lifted. In Europe, the Dutch have a very organized trade in bulbs of tulip, gladiolus, iris, *etc.*

Prolonging Cut-Flower Life

If we are to succeed in the cut flower trade, research must be directed towards prolonging the life of cut flowers and scientific packaging so that the flowers may reach the consuming centres in garden-fresh condition. The specification for exporting cut flowers is that the intervening period between cutting the bloom in the field in India and its arrival in the European market should not be more than 36 to 48 hours. The technique of pre-cooling of the bud before shipment for prolonging its life should be mastered for each flower

Flowers dried in their natural colours offer U.S.A., and Japan. Australia has specialized in this trade and is exporting these to Japan Borax is used in drying flowers and foliage. The flowers are dried by keeping them in standing position in a container and covering the whole flower carefully with borax. The flowers remain in borax till all the moisture is absorbed. The importance of gardening is also not understood in India. The people should be educated to realize the importance of gardening, by providing good parks in cities for a large population to relax and enjoy the beauty of nature and gardens.

Landscape Gardening

The importance of gardening is also not understood in India. The people should be educated to realize the importance of gardening by providing good parks and gardens in cities for a large population to relax and enjoy the beauty of nature.

Bio-aesthetic Planning

The term bio-aesthetic planning, a concept of Prof. Lancelot Hogben, means the proper utilization of the available flora and fauna in the beautification of the surroundings. The aim behind this concept is to plant ornamental flowering trees along roads, in parks, house compounds, public places, and also to develop national parks where non-carnivorous animals and beautiful birds will find sanctuary along with beautiful flora. This term also includes landscape gardening though in a wider sense. Bio-aesthetic planning can be extended to the entire country. The bio-aesthetic planner can be described as a master artist who uses the whole country as his canvas and his plants are the rich colours of red, blue, orange, and white of the different flowers. It is said that the un-touched nature is quite monotonous. It is only with a touch of bio-aesthetic planning that the countryside and other places will look not only natural but pleasing too to the eye.

Bio-aesthetic planning should run hand-in-hand with town planning. Our new townships should not be allowed to grow as mushrooms in dung-heaps as our older towns are. The roads in towns and cities should be broad, planted with flowering and shade trees, and there should be spacious parks along with conservatories for harmless animals and birds.

The older congested cities and towns also should be retrieved from their present state by bio-aesthetic planning. One way of doing this is to acquire by compulsion the ugly areas of the towns in some centrally located pockets and to convert them into parks. The displaced persons may be accommodated in multi-storey buildings which occupy less land. But many planners are against vertical growth in our towns; a compromise must be found somewhere.

Air Pollution

Air pollution, one of the most-talked-about problems in the present age, has reached disturbing proportions in some of the largest cities of the world and also in some of the metropolitan cities in India. The unplanned growth of the cities has resulted in congestion of houses, factories in residential areas, and proliferation of motor vehicles. The smoke from the coal ovens (chulas) of the residential houses, the dust and the smoke from the grinding mills and chimneys of the factories, and the exhaust fumes from the motor vehicles — all add to the pollution of our respiratory tract, cancer, and many other ailments. Unless something is urgently done the health of our citizens may deteriorate rapidly.

The role of open spaces such as parks and of living plants in checking air pollution is well known. The parks are considered as the lungs of a city. The barrier of trees checks noise pollution, dust pollution, and air pollution.

Human Welfare

The role of landscape gardening in human welfare cannot be overlooked. Even in an under-developed country as India, people do not live by bread alone. They also need some finer things of life. It is a great tragedy that most of our children in big cities do not have any open space to play and to see colourful flowers, birds, and butterflies. It is the moral duty of our government, through the municipalities, corporations, and such other bodies to provide the citizens with spacious parks having beautiful trees and flowers where they can relax, find peace of mind, and breathe fresh air after a day's hard work. The children will also be able to groups of youngsters play football, cricket, or hockey in the by-lanes in the absence of playgrounds and parks. Such a state of affairs should not be allowed to go on indefinitely. The wealth of any nation is linked with the health of its people. Unless we can ensure the healthy development of our citizens, especially the younger generation, by providing for them open breathing places through bio-aesthetic planning and landscape gardening, we cannot expect to build up a healthy society and a prosperous nation.

Recent Status of Floriculture in India

- Leading flower product exporter- Netherland
- Leading flower product importer- Germany
- Country having large market of cut flower – Germany
- State having maximum area under – Karnataka
- State having maximum production – Tamil Nadu
- Largest importer of floriculture product from India - USA (27 per cent)
- Maximum cut flowers production in India – West Bengal
- Flower cover maximum area in India - Jasmine
- Japanese flower arrangement also called - Ikebana
- Largest producer of loose flower - India

Favourable Factors for Floriculture Industry in India

Climate

The varied climate in India makes it most suitable for growing flowers in one part or the other part so that flowers can be produced the year round. For example, roses, gladiolus, carnation, chrysanthemum, *etc.,* are successfully grown in northern plains during winter months, whereas, summer months flower can be grown in hills. Growing of wide range of flowers is also possible in different parts of India. India being native of many flowers opens up new view for exploiting the commercial growing of rare flowers for export.

Soil

The soil are different climatic zones is very favourable for growing different type of flowers. Suitable soils are used for the good cultivation of the flowers.

Labour

Comparatively cheap and easily available labour in India is another favourable factors for growing flowers at cheaper rate.

Transport and Storage

Flower after harvesting the marketing for highly benefit market. The storage of the flower at favourable condition for long duration and marketing.

Chapter 3

History of Gardening in India

The history of gardening in India is as old as its civilization. The first recorded history of civilization in India dates back 2500 B.C. before the Aryan came. The civilization prospered in Punjab, Rajasthan, Haryana and Gujarat. The first evidence of an ornamental plant, the pipal (*Ficus religiosa*), comes from a seal from Mohenjo-daro, of the third millennium B.C. another seal from Harappa of the same period depicts a tree similar to that of a weeping willow (*Salix babylonica*).

The origin of planet earth took place about 500 million year ago whether the life first appears in the form of blood or chlorophyll is still not known clearly. The simpler form of flora and fauna, passed through evolutionary process and development in to complex system. Dominance of flowering plants started during cenzoic age 60-70 million years. According to N.I. Vavilov Indo-Malaya Region is one of the important centers of origin of plant. Many flowering plants are strictly native of India, *e.g.*, orchids, balsam, muskrose, tulip, amaltas, kadam, kachanar, *etc.* The history of systematic gardening in India is as old as civilization of Indus of Harappan. Which existed between 2500 B.C. and 1750 B.C. people where living in well planed dwelling. A road cuts across one another almost at right angle. The tree were associated with the human life in many ways is quite clear from the seals available of that period. The picture of God is presented in the middle of branch of pipal. Besides pipal, mango and willow have also been depicted on the seals. Harappan poets were generally decorated with the design of tree. The cult of tree and animal worship was at its peak. The pipal (*Ficus religiosa*) and banyan (*Ficus benghalensis*) trees served mankind and fauna in many ways. That is why till today these trees are being worshipped. In every village the tree especially the pipal and banyan are planted for worship as well as for shade.

The Aryan civilization dates back to about 1600 B.C. The Aryan of the Vedic period were great lovers of trees and flowers. The lotus has been mentioned frequently in the Sanskrit scripture of Vedic times. Land was covered with the trees of palash laden with scarlet flowers. Lotus being a native of India was found everywhere. The Aryans were literary people and brought with them the four Vedas *i.e.* Rig Veda, Atharva Veda, Yuzur Veda and Sam Veda and the Puranas. The Atharva Veda mentions the pipal or banyan. Possibly the Rig Veda also refers to this tree when it describes a tree with fair foliage. The use of flowers in religious and social ceremonies was started by Aryans. The forest of Sal (*Shorea robusta*) and flame of the forest (*Butea monosperma*) were admired by Aryans. The Lord Vishnu and goddess Luxmi were associated with Lotus.

The association of different trees with the life of Lord Buddha is well known. The Buddha was born in 563 B.C. under the tree of Asoka at Lumbani in the grove of Sal. In this grove the trees of palash were also there, hence, all the three trees were associated with the birth of Buddha. The great Chinese traveler, Hiuen Tsang, who comes to India (630 A.D.), confirmed that the Buddha was born under the Asoka tree. The Buddha attained his enlightenment under a Pipal tree, spread his new teaching under shady banyans and mango trees and breathed his last in a Sal grove.

The great Emperor Asoka (264 B.C.) adopted arboriculture as one of his state policies. He encourages the planting of tree alongside the roads (Avenue tree). His son Prince Mahendra took a sampling of the great bodhi tree (*Ficus religiosa*) and planted it at Anuradhapur in Sri Lanka (250 B.C.).

- **Gardening during Epic-era-** Ramayan (written by Valmiki and Tulsidas) and Mahabharat (written by Saint Vyasa)
- **Gardening during Mughal-era-** Babber (1540 A.D.)
- **British period-** King Hyder Ali (1782 A.D.) and Tipu Sultan (1789 and 1792 A.D.)
- **Post-independent period-** Indian Government (1947)

The famous poet "Bana-Bhatt" describe a number of flowering plants including the Banyan, Sal, Palash (Flame of forest), Kadam, Ashoka and the Indian coral in history in his famous book "Harsh-Charitra", Vatsayana (300- 400 A.D.) in his book 'Kamsutra' given a glimpse of joyfull civic life of that period. There is also description of four types of garden during that period.

- Garden meant for the enjoyment of the royal couples called **Premoudyan.**
- Place where the kings played chess enjoyed the dances of the maids and jokes of the court jesters named **Udyan.**
- The garden where high-placed persons in the king's court enjoyed life with courtesans called **Vrikha-vatika.**
- **Nandanvana** was associated to Lord-Krishna.

The post-independence development in India was characterized by the creation of several public parks and gardens of formal and informal style. During 1960-61, the Government of India has invited a Japanese garden expert, Prof. K. Mori for advice to help in creation of Japanese gardens in India. These gardens were developed in the Roshanara Garden, Delhi and Buddha Jayanti Park, New Delhi.

In order to make growing of flowers into a successful proposition, it should be based on scientific research. For this the Indian Council of Agricultural Research is playing a vital role and has started a co-ordinate approach. Availability of technically trained persons is also very important factor for making a programme successful. Till 1971, no university was provided education up-to M.Sc. level in floriculture and landscaping. Punjab Agriculture University, Ludhiana, was the first Indian University to start imparting education in floriculture and landscaping. Now the facility is available at other university like Solan, Coimbatore, Kalyani, Pune, Kanpur and I.A.R.I., *etc.*

FAMOUS GARDEN OF INDIA

- ☆ Botanical Garden (Forest Research Institute)- Dehradoon, Uttarakhand.
- ☆ Vrindavan Garden – Mysore, Karnataka.
- ☆ Buddha Jayanti Park- New Delhi.
- ☆ Chashma-e-Shahi Garden- Srinagar (J & K).
- ☆ Government Botanical Garden- Octacamund (Tamil Nadu).
- ☆ Indian Botanical Garden (Royal Botanical Garden)-Sibpur (West Bengal).
- ☆ Lal Bagh Garden-Bengluru (Karnataka).
- ☆ Nishat Bagh Garden- Srinagar (J & K).
- ☆ Pinjore Garden- Harayana.
- ☆ Rambagh (Aram Bagh) Garden- Agra (U.P.).
- ☆ Rastrapati Bhavan Garden- New Delhi.
- ☆ Rock Garden- Chandigarh.
- ☆ Roshanara Park- New Delhi.
- ☆ Shalimar Garden- Srinagar (J & K).
- ☆ Sayaji Park- Vadodra (Gujarat).
- ☆ University Botanical Research Institute- Coimbatore (T.N.).

Chapter 4

Principle of Gardening

The planning of a garden is an art. It is said that a poet is born, and in that sense a garden architect is also born. But that does not mean that a garden architect has nothing to learn. On the contrary he should learn enough of geology, geography, garden history, styles of gardening, and above all should have a profound knowledge about plants. It is imperative to know at the outset what is meant by landscape gardening. A landscape may be defined as any area, either big or small, on which it is possible or desirable to mould view or a design, whereas landscape gardening may be described as the application of garden forms, methods, and materials with a view to improving the landscape. The art of designing is known as "landscape architecture," although the older term landscape gardening is also popular.

Before discussing the principles of gardening in detail, it will be worthwhile to make some useful suggestions. The first and the foremost thing are not to imitate another garden which has secured a prize in a competition. One has to develop one's own design giving due consideration to the local conditions. One more mistake which is commonly made is to plant more specimens than a garden can accommodate causing overcrowding. In a landscape garden any difference in levels has to be taken advantage of, but in a perfectly flat land it will be costly to create artificial undulations. In each garden the care should be at least one feature of there be a second these two should harmonize with each other. Before planning a design one must be sure for what is purpose of the garden utility or beauty or both.

Landscaping is an art of beautifying a piece of land or a landscape with planting material non-living material in order to create a picturesque effect or to imitate natural. Landscaping makes a place more peaceful, beautiful, appealing of pleasing, where people can rest and enjoy with their family and friends, *etc.* Landscape gardening is landscaping of a garden, more or less landscape gardening and painting on a canvas or paper sheet is similar. In both cases colours are used to form a picture, only difference is that painting is non-living 2D image and garden is collection of living plants with non living things in 3D form. It is not necessary and also possible to follow all the principles at a time in a landscape plan but there are basic principles one should follow during landscape planning, which are discussed below.

1. Simplicity

The plan must be simple. It should not contain too many themes or overcrowding of plants or other garden ornaments, as this creates confusion and hinders in getting the theme and also distract the viewers.

2. Unity

It increases aesthetic beauty of the garden. It can be expressed by harmonious placement of garden features in proper way. It can be created by repetition of single component or by blending different component in harmony in such a way that every feature expresses the total effect independently.

3. Harmony

It is an overall effect of various features, style and colour schemes of the total scene. It is arranging different things in garden in such a way that it creates relation between all and makes them all look one.

4. Balance

Balance in a design is an essential part for unity, harmony and proportion of design. Therefore, correct balance in different planting materials like annuals, shrubs and trees should be maintained. Balance in size, shape, colour and form of building and different plant should be maintained.

5. Proportion

In this every feature should be in relation with total space available for the garden. The space provided for lawn, paths, herbaceous borders, shrubbery, tree, building and other garden object should be in a right proportion. It should be seen in Mughal garden like size of building, tree, lawn, paths, *etc.*

6. Scale

Scaling of components, size, shape and form unites the different component of design, create harmony and incorporate balance and proportion in design. In general in a design, area can be divided in following manner which may vary with personal choice and type utility. Where on the lawn (25-30 per cent), paths (15-20 per cent), herbaceous border (8-10 per cent), shrubbery (12-15 per cent), trees (15 per cent) and building (25-30 per cent).

7. Accent/Focal Point

Accent is a centre of attraction which is generally an architectural feature focused as a point of interest. The focal point attracts viewers to a point from where a single view can be seen and from where the whole plan could be understood and all components looks one which avoids monotonous view. The hidden focal point is called as vista. Mostly objects like fountains, tree and statues, *etc.* are used. In English gardens statues are used as focal point. It may be anything, a flowering tree, shrubbery, focal clock, fountains, lakes, *etc.*

8. Rhythm

Repetition of same object or objects at equidistance is called rhythm. It can be created through the shapes, progression of sizes or continuous line movement. Rhythm creates movement of eye. For example in Mughal gardens trees of single species of equal height and shape, fountain and water canals are planted to create this effect. In present day situation, light are used for this purpose.

Initial Approach

In theory, everyone would like to have a perfect plot of land, but in actual practice the plot available for gardening, in three out of five cases, either will not be in a good or the shape and size will not be ideal. Whatever may be the case, one should not throw hands up in despair even if the site appears to be hopeless. A good designer is one who will make best use of such a site. As has already been stated, land with natural undulations should never be levelled, but rather the differences in levels should be utilized with advancing, though may not look artistic is essential in any garden. But the fencing should be such that it looks naturals far as practicable and it should not obstruct any natural view. For example, if there is natural forest scenery or a hillock just outside the boundary it should be incorporated in the garden design in a thoughtful manner so that it appears to be part of the garden.

A building or machinery, once it is complete, takes a final shape. But with living plants it is impossible to visualize how a garden design will look like in the long run when the plants attain full size, even though the design may have been completed. For example, a tree which is normally expected to grow to a height of 10 m, remains dwarf for some reason or other thus jeopardizing all calculations. One more difference with inanimate objects and living plants is that while the final design for a building or machinery can be completed on paper this cannot be so with living plants, and as such each design should be adaptable. The man on the spot who will actually implement the design should be given enough scope to change it to adapt to the local needs and personal taste of the owner and/or designer. It is only the formal gardens, not having any trees that can be drawn on paper and implemented without any appreciable change.

The other terms and elements used in landscape designs are briefly discussed below.

Axis

This is an imaginary line in any garden around which the garden is created striking and balance. In a formal garden, the central line is the axis. At the end of an axis, generally there will be a focal point, although another architectural feature such as bird-bath or sundial can also be erected

Mass Effect

The use of one general form of plant material in large numbers in one place is done to have mass effect. To see that such mass arrangements do not become monotonous, the sizes of masses should be varied.

Unity

Unity in a garden is very important as when this is achieved it will improve the artistic look of the garden. Unity has to be achieved from various angles. First, the unity of style, feeling, and function between the house and the garden has to be achieved. Secondly, the different components of the gardens should merge harmoniously with each other. The aim is to give the visitor an overall impression of the garden rather than blowing up some special features. The last point, which is also very important, is to achieve some harmony between the landscape outside and the garden. A garden laid out in complete defiance of the local conditions may look exotic, but is not a successful garden. As for example, cacti planted in a seashore garden is completely out of place as these are inhabitants of dry localities.

Space/Volume

The aim of every garden design should be such that the garden should appear larger than its actual size. One way of achieving this is to keep vast open spaces, preferably under lawn and restrict the plantings in the periphery, normally avoiding any planting in the centre. But if any planting has to be done in the centre the branches at a higher level on the trunk (or the lower branches are removed) and not a bushy shrub. Such planting will not obstruct the view or make the garden appear smaller than its size. Another way of creating an illusion of distance, which was in vogue in older a yes but not practised much in recent times, is to make lines converge slightly at a distance. For example, paths in a garden are gradually narrowed or the sizes of the farthest trees diminish in size. But such tricks will give a reverse impression from the other end. Another suggestion to create the illusion of more space in a large public garden is to alternate large. A large open space planted haphazardly all over with trees looks smaller than its size. The technique of creating an illusion of more space is also referred to as forced perspective

Divisional Lines

In a landscape garden, there should not be any hard and fast divisional lines. But there is the necessity of dividing or rather screening a compost pit or a alts quarter or a vegetable garden from the rest of the garden. In fact areas under lawn, gravel, stone or cement path, and shrubbery border have their natural divisional lines from its immediate neighbour though these are not discreet. This is what is exactly needed. The divisional lines should be artistic with gentle curves and these should also be useful. Above all these lines should harmonize with one another.

Proportion and Scale

Proportion in a garden may be defined as a definite relationship between masses. For example, a rectangle having a ratio of 5 : 8 is considered to be of pleasing proportion. As this ratio comes down the form looks neither a square nor rectangle, and the design becomes undesirable. There is no set rule as regards scale or proportion in a garden. But a simple rule is that a design should look pleasant. It is better to have an *ad hoc* design first and then try it out on the actual spot. If the design looks appealing as well as pleasing, it is implemented. When a shrubbery border has to be planted the outer design is marked by arranging a rubber hose or thick wet rope in different designs on the spot and the one which looks best is adopted. Then sticks of different heights, representing the various shrubs, are planted in various positions and by the method of permutation and combination the most proportionate looking arrangement is adopted.

It will be worthwhile to give a few more examples to illustrate the importance of scale in a garden. A narrow step leading from a wide terrace is completely out of scale. The steps in a garden, should not only be broader than those inside the house but should, have deep treads (the stepping) and low risers also. This means the steps are spaced wider, making climbing easier and pleasant. Moreover, a very wide flight of steps dividing two lawn areas, at different levels, make the transition clumsy and inconspicuous. The common practice of laying out a small rockery at the base of a large tree with small thorny specimens looks not only ugly but is also out of scale and proportion under the large canopy of the tree. A tiny pool in the midst of a large lawn also looks disproportionate.

Texture

The surface character of a garden unit is referred to as texture. The texture of the ground, the leaves of a tree or shrub will all determine the overall effect of the garden. The texture of rugged looking ground can be improved to an appreciable extent by laying meticulously chosen small pebbles from the riverbeds, if establishing a lawn is out of the question. A gulmohar is a fine textured tree when in full leaf, whereas *Spathodea companulata* is a coarse textured tree. The placement of all these various textures with harmony and contrast has to be achieved to get the ultimate desirable effect.

Time and Light

In a category of time in a gardens, first comes the daily time, which provides different quantities and qualities of light during the course of the day. As the morning sun is vital for all flowers, the designer has to take this into account while planning. In most parts of India the garden design should be planned in such a way that in the afternoon it is possible to sit in a shaded place from where the best part of the garden should be visible. The second type of time is the seasonal changes in the year.

Mobility

In a temperate country, the garden changes colour very sharply and contrastingly from one season to the other thus symbolizing mobility or movement. As for example, many trees in the temperate regions attire themselves with wonderful due to the changes in their leaf colour in the autumn. Then suddenly in the winter leaves fall and everything goes to rest bringing an atmosphere of melancholy and dullness all around. Again in the spring the plants spring back to life with the appearance of the new leaves. In most parts of Tropical India, though these contrasting changes cannot be achieved, it is possible to bring in some subtle changes, such as Bengal, Indian Almond (*Terminalia catappa*) which change its leaf colour into striking red twice annually before falling or *Lagerstroemia flos reginae* which also changes the colour of the leaves to coppery shade in the autumn before shedding, or *Madhuca indica* and *Ficus religiosa*, the new foliage of these appearing as coppery red in the spring, should be planted in some parts of the garden.

Style

Lastly, one has to decide about the style to be adopted for one's garden, broadly speaking, every garden lover has to invent his own style of gardening commensurate with his budget, taste, and the nature of the site. But a man can develop his own design only when he studies carefully all the great garden styles of the world and grasps the underlying principles in them. There is no doubt that persons not having enough specialized knowledge will commit mistakes; nevertheless, one should not get deterred by this fact. One word of caution to a novice gardener is that he should not get used to his mistakes, but critically assess every feature and try to correct the mistakes and improve upon the design with the acquiring of new knowledge through experience and learning from others.

Colour

Colours are the most essential element in a landscape design. This imparts beauty and cut boredom, monopoly of green colour. Colours are divided into three groups.

Primary colour: Red, yellow, blue.

Secondary colour: Green, violet (purple), orange.

Tertiary colour: Mixture of primary and secondary colours.

White, grey, black and silver colours are considered neutral.

Colours include three terms namely hue or chroma, value and intensity. Hue or chroma refers to the relative purity or strength of the colour. Value determines how light or dark the colour is, whereas, intensity refers to how bright or dull it is.

Habit

Habit of planting material (straight, globular, columnar, bushy shape of crown) attracts viewer's eye movement in different directions. Straight growing plants move viewer's eyes towards up where sky and tall plants jointly give a scenery view, adding sky in landscape design. Globular plants move eyes in horizontal direction which relates earth and other small components. Habit of plants play an important role in viewer's eye movement and help the viewers by guiding what landscaper wants to show or the beauty of his design. Habit can be compared with silent direction signs, indicator or guide.

Chapter 5

Lawn

Lawn is a natural green carpet, and it is important feature of a landscape and a garden without a lawn is not considered complete. Lawn is an area of land closed mowed grasses and it is primarily developed for aesthetic and recreational purpose. A lawn is an integral part of a garden and landscape besides aesthetic and recreational purpose and several other purposes. Lawn provides a place for taking rest after tiring job of the day, holding parties, social functions, passive and active recreation. Distinguishing characteristic of turf grasses is the ability to with stand close mowing and still providing a functional dense healthy ground cover. Lawn is also called as heart of garden. Lawn can be called as turf pitch a field, sod, yard, are green depending upon the plantation usage continent. Lawn is a ground cover of perennial grass, ditch, persists enclose mowing and require proper management practice.

J.B. Olcoth designed first experiment in lawn grass at Monchester in USA in the 19th century. Festuca and Agrostis species were used to make turf (Lawn) during 1880 in U.K. In the early 19th century lawn become wild spread out side of parks, golf course. Nowadays in many metro cities of India lawn has become a culture of the passion of many homes.

Selection of Grass

There are mainly two types of grasses are used for planting lawn.

1. Bermuda Grass (*Cynodon dactylon*)

It is commonly called as doob or hariali. This is very commonly used for planting lawn due to its fast, growth, hardiness, less water requirement and response to frequent mowing. This makes an excellent turf. It is highly suitable for large area and a playground on account of its tolerant to water and has good recuperation habit.

Variety of burmuda grasses- Kalkatiya, Hariyali, Selection 1, 5, 8, Palna, Panam, *etc.*

2. Korean Grass (*Zoysia japonica*)

It is native of Japan, Korean and Philippine island. This kind of grass is recently introduced in India and has become very popular in short span of time due to its velvety growth and more cold tolerance. It makes cushion like turf. This is highly suitable for smaller areas and home lawns.

Types of Lawn Grasses

Warm Season

Warm season grasses grow luxuriantly in warm climate they starts growth at the temperature above 10°C and grow fastest when temperature are between 25-35°C and they are comparatively harder than cool season grasses. Warm season grasses are more popular for making lawns, they require less care and maintenance, *e.g.*, Doob grass/barmooda grass (*Cynodon dactylon*), Korean grass (*Zyosia japonica*), Buffalo grass (*Buchloe dactyloides*), Carpet grass (*Axonopus affinis*) and Bahia grass (*Paspalum notatum*), *etc.*

Cool Season

Cool season grasses start growth at 5°C and grow at their fastest rate when temperature prefer between 10-15°C. Some important cool season lawn grasses are Tall fescue (*Fesuca arundinacea*), Chewing fescue (*Festuca rubra*), Annual rye grass (*Lolium multiflorum*), Creeping bent grass (*Agrostis palustris*), and Kentucky blue grass (*Poa pratensis*), *etc.*

Purpose of a Lawn

- It is an important element in garden.
- It leads to unity in garden design.
- It is natural green carpet and is the carpet floor of outdoor room.
- It is a heart of garden and the center for social life.
- It gives restful appearance to the eyes through its green outlook all the time.
- Prevent soil erosion.
- Enhance ground water recharge.

Characteristic of Lawn Grass

- Look fresh and green throughout the year.

- ☆ Cold and drought resistance.
- ☆ It should not be patchy.
- ☆ Free from attach of disease and insect.
- ☆ Quick growing.
- ☆ Should not give foul or bad odour.

Preparation of Soil

1. Dig soil up to 45cm depth and expose sun May and June.
2. Turn soil 2-3 times remove stone rocks and break big clouds
3. Spread 10-15 cm thick layer of well rotten weed free FYM and thoroughly mixed in soil.
4. Sandy loam soil is best for lawn grasses growing which pH ranges between 5-6.
5. Irrigation the field thoroughly and allow all weeds to germinate.
6. Remove the entire weed along with roots manually or spread non-selective type of herbicide, like Paraquat or Gramaxone at the rate 1-1.5 liter in about 800-1000 liter.

Site for Planting Lawn

South, South-East, South-West open and sunny place for most part of the day with adequate water availability.

Planting Methods of Lawn

Following are different methods of planting lawns:

1. Seeding Method

- ☆ This method is common to grow cool season lawn grasses.
- ☆ About 25-30 kg seed/ha is mixed in 200-500 kg sand and saw dust and broadcast evenly in prepared field.
- ☆ Do light rolling.
- ☆ Sprinkle water regularly until seedling emergence.
- ☆ Less labour is required but lawn is not even.

2. Dibbling Method

- ☆ A small bunch of grass alone with roots and little stem is taken.
- ☆ Planting is done at spacing of 10 cm apart both row to row and plant to plant.
- ☆ Do regular watering until establishment.
- ☆ It is done in June to September.

☆ Lawn develop by this method is quick uniform and with more labour and cost.

3. Sodding Method

☆ Sodding is expensive than others vegetative propagation method.

☆ This method is recommended where quick cover is required.

☆ The sod should not be laid when the day time temperature excess 32°C for an extended period.

4. Sprigging Method

☆ Grass root along with little stem are chopper into small pieces.

☆ Spread this over prepared field during rainy season.

☆ Do small raking to mix grass in soil.

☆ Do light rolling.

☆ Do laboural watering with sprayer.

☆ Do moving after 70-80 days.

5. Turfing Method

☆ Small pieces of well prepared lawn or turf are cut into square or rectangular shape.

☆ Preferable plants are planted in polythene seat.

☆ Fix these in thoroughly prepared field.

☆ Do heavy rolling.

☆ Lawn prepared is clean and weeds free.

☆ Quickest method of lawn raising.

6. Plugging Method

☆ The planting of 5-10 cm diameter square, circular or plucked shape piece of sod at regular interval is called plugging.

☆ Care should be taken that plugs or well watered but must not be soggy (muddy).

☆ There is 10 times more planting than sprinkling.

☆ Quick establishment of lawn, it is advised to plant plugs closely.

7. Stolonizing Method

☆ Stolonizing is the broad casting of stolons on the soil surface.

☆ Covering by top dressing or pressing into the soil.

☆ This method requires more planting materials.

Management of Lawn

Mowing and Rolling

The newly planted lawn is not mowed till has established well. In the beginning the grass should be trimmed with the help of grass cutter (Mower). It is the cutting of lawn grass for maintaining its activeness and for maximum utility. Mower was discover by "Edwing Budding" in 1830 in England. Light rolling should be done on dry ground to suppress the upright growth to anchor the grass firmly in the soil and to keep the level of ground. Interval of mowing of an established lawn depending upon season. In rainy and winter season mowing is required at an interval of 7-10 days. During spring season mowing is required at the interval of 15 days whereas during summer it is done at monthly interval.

Manures and Fertilizers

The regular application of fertilizers keeps the grass to grow luxuriantly and maintains the lush green colour of the lawn. Sun hemp is very good green manure before lawn planting. In 30 meter square lawn 3-5q. FYM, 10-20kg lime and 10- 20 kg SSP are required. Broadcast a mixture of 50-60gm/m^2 or 1.5kg/30 m^2 (2 CN: 1 SSP: 1 K_2SO_4) twice in Feb-March and Aug-Sept and spray of urea at 0.3 per cent.

Irrigation

Water requirement of lawn depend upon season, type of soil, grass used, weather and planet. Grasses are surface feeder and hence adequate watering should be done frequently. It should be done before wilting due to internal water stress. Increased watering interval results in deeper root development, then by decreasing water requirement. Irrigation required up to 5-15 cm on depth at 8-10 days interval is ideal as frequent light watering is harmful. Korean grass needs more frequent watering than Kalkatiya grass.

Weeding

Regular mowing checks weed growth by removing upper portion of the weeds and starvation of root. But still there may be large population of weeds which grows fast and needs removal. For controlling broad leaved weed spread 2-4-D at 0.005 per cent and for narrow leaved weed spray Atrazene @ 1.5 kg/ha in 1000 litter. In lawn of Korean grass spray Benefine or Sylvex at 0.1 per cent.

Scraping of Lawn

It is done to renovate the old lawn when it becomes old and grass become compact. After 3-4 years during summer month, *i.e.,* during June, lawn should be scraped completely with the help of khurpa and raking should be done both ways. Before the starts of rains top dressing of mixture consisting of garden soil, sand and sieved leaf mould (1:2:1) should be applied to cover the upper 3-5cm.

Disease

- ☆ Damping off, root rot, gray leaf mould, leaf spot, powdery mildew, fairy ring are most important disease of lawn.
- ☆ It can be controlled by Bordeaux mixture 4 : 4 : 50 at fortnightly interval.
- ☆ Dithene-M-45 (0.1 per cent) + Bavistin (0.1 per cent) at fortnightly interval.

Insect

- ☆ Termite (Controlled by 0.03 per cent Chloropyrophase), cut worm (Control by Spinopad at 0.05 per cent) and root grub (Controlled by Phorate 10-12 kg).

Silent Point for Lawn

- ☆ Avoid water standing in rainy season.
- ☆ Remove all dead or dry leaves falling during autumn season.
- ☆ Do raking in lawn twice, once before rain and 2nd after rains.
- ☆ Do thinning when required.

Chapter 6

Hedges, Edges and Topiary

HEDGES

When shrub is planted on boundary for fencing, it is also called as hedge. In other word shrubs or trees planted at regular intervals to form a continuous screen is called a hedge. It may be ornamental or protective or both. Hedge also become an important feature of formal layout to serve various function as screening of the area out building, tennis court, vegetable garden, hardening of unwanted place, *etc.* All the shrubs cannot be planted for making an ideal hedge due to different nature of growth.

For selecting an ideal shrub for a hedge, it should have following characteristics.

- ☆ It should have thick texture and quick growth.
- ☆ It should stand trimming to shape.
- ☆ It should be easily propagated through seeds or cutting.
- ☆ It should withstand drought condition.
- ☆ It should not attract reptile's animals.

Classification of Hedges

Hedge are planted to protect the area to avoid the trespassing to men or animals or to beautify the boundary of different therefore according to purpose hedge can be classified as–

- ☆ Tall protective hedges.
- ☆ Dwarf protective hedges.
- ☆ Tall ornamental hedges.
- ☆ Dwarf ornamental hedges.

1. **Tall protective hedges-** The height is about 1-3 m and growth is very dense along with thorns, *e.g., Inga dulcis, Carissa carendus* and *Bougainvillea* species, *etc.*
2. **Dwarf protective hedges-** Dwarf shrubs grow about 1m and have thorns and protective in natures, *e.g.,* Euphorbia (*Euphorbia bojori*), Nagphani (*Opuntia* species), Agave (*Agave* species) and Davils tree (*Pendilanthus* species), *etc.*
3. **Tall ornamental hedges-** Shrubs grow 1-3 m tall and have attractive foliage plant produce colour full flower are also, *e.g.,* Heena (*Lawsonia alba*), Duranta (*Duranta plumier*), Rukmani (*Murraya panniculata*), China rose (*Hibiscus rosasinensis*) and Fire bush (*Hamelia patens*), *etc.*
4. **Dwarf ornamental hedges-** Height of dwarf hedge is about 1m and plants are very attractive, *e.g.,* Acalypha (*Acalypha* species), Wild jasmine (*Clerodendron inerme*), Lantanas (*Lantana species*) and *Thunbergia erecta, etc.*

Growing of Hedges

Purpose

A garden hedge can serve the purpose of a compound wall, give shelter from strong gails, ensure privacy, *i.e.,* serve the purpose of a screen from a background for a floral display such as herbaceous border, as a part of the garden on its own merit, separate one component of a garden from the other (vegetable garden from the flower garden) and screen the ugly and unwanted spots such as manure pits, lavatory, servants quarters, *etc.,* in the garden.

Criteria for Selection a Hedges Plant

In a garden a hedge is planted with two motives: (a) protective, which means protection against theft, trespass, wind, *etc.,* and (b) for ornamental purposes or screening. For the first category a hedge plant should have the following characteristics- quick growing, hardy, including drought resistant characters, thorny, dense, should responded to frequent pruning and clipping and can be raised quickly by seeds or cutting.

Preparation of Soil

Hedge plants can be grown in various kinds of soil. For better growth of plants soil should be deep, well drained and fertile therefore, planning and preparation of ground should be done carefully. For making a good growth of hedge a trench about 60-75 cm deep and 30-60 cm width should be dug up and left exposed for a fortnight or so before planting in order to destroy harmful micro flora and insects by scorching sun.

Planting Time

The most suitable time of planting hedge is rainy season. In the beginning of monsoon, weather condition are hot and humid. Planting can also be done during Feb-March depending upon the availability of regular supply of water and planting material.

Planting Method

Rooted cutting or seed are planted. Under Indian conditions the planting of hedge is undertaken at the beginning of the monsoon season in June-July. A hedge is started by sowing seeds or by putting cutting *in situ* or by planting rooted cutting, generally in double rows. The distance of planting for tall hedge should be about 60-90 cm and for dwarf hedge it should be about 20-30 cm. The planting should be done by triangular system, show that it makes a good dense hedge.

Maintenance

Hedge does not like weed growth at the base and these should be uprooted as and when they appear. Keeping reserve plants are essential, otherwise the casualties are to be replaced by seeds or new cutting, again the uniformity in growth will be lost.

Irrigation

Water requirement of hedge depends upon climate and soil types. During rainy season generally not irrigation required for hedge planting and during winter season generally water required once in 10-15 days and at weekly intervals during summer season.

Manuring

Manuring is one aspect which receives least attention by the gardeners. One may be inclined to believe that manuring of hedge is a sheer waste. Once in a year the hedge should receive before the rains well rotten cow-dung at the rate of 4 kg per running metre. It should be well incorporated in the soil. Fertilizers can also be added twice a year at the rate of 30 g each of super phosphate, bonemeal and ammonium sulphate per metre row of hedge.

Trimming

When the hedge plants attain a height of 15 cm height with the topped back to 10 cm height with the garden shears topping is done many times till desire height is achieved. Later on hedge can be trimmed to different shape. The top of hedge can be kept flat, wavy, black of square or any shape. In rainy season frequent trimming is required because plant make faster growth than winter or summer season.

Clipping and Pruning

It is most important to keep the hedge neat and in good shape. They should be done at regular intervals and at no stage the top growth should exceed 15-20 cm in length.

EDGES

When low growing perennial plants are grown on the border of plots or beds, they are called as edge plant or an edge. These plants hardly grow up to 20-30cm. In garden hedges are also planted around rockery, big tree, alongside walks pathway and to divide the area. Like hedge they also become a part of formal or informal landscape designs. The hedge plants are usually propagated through terminal cutting during rainy season and stand against trimming, *e.g., Eupatorium cannabium, Justicia species, Iresin lindenii* and *Alternanthera versicolour, etc.*

Edging

The term edging can be defined as any material of any description which is employed in garden for dividing flower bed borders, *etc.* from road wall or path for demarcating spaces allotted for particular purpose.

Kinds of Edging

1. **Living edging-** Colourful flower
   a. **Foliage plants-** *Alternanthera tricolor, Aspidistra* sps., *Coleus blumei, Iresine* sps. *and Pilea muscosa, etc.*
   b. **Flowering plants-** Sweet allysum (*Allysum mauritinum*), Candytuft (*Eberis amara*) and Marigold (*Tagets* sps.), *etc.*
2. **Mechanical edging-** Brick in vertical and horizontal

TOPIARY

It is an art of training plants into different shape, *i.e.,* of birds, animals, domes, umbrellas, *etc.* Topiary was first introduced in Roman garden in the first century and later revived in 12th century in the monastic garden. For simple geometric topiary shape frame is not required and the plant can be trained free hand. However, for larger shape, iron frame is required to trained the plant on it. It is an old art and nowadays. It is becoming common in city park to provide passive, recreation to the visitors specially children. The ideal plant should be quick growing with small foliage and stands against trimming which is done frequently.

Suitable plants like *Clerodendron inermi, Cupressus macrocarpa, Duranta plumier, Bougainvillea species* and P*utranjiva roxburghii, etc.*

Training frames are employed for making topiary. The frames are generally made by soft steel rods or galvanized wire frame. Making frames true to shape is highly technical and artistic for true depiction of figures. The figures of lady with a pitcher, elephant, giraffe, camel, ox, monkey, birds, peacock, farmer and bullocks, umbrella, *etc.* are liked by everybody. For picturesque topiary, there should be a broad base or on platform of plants. For large animals 2-4 plants are planted at different point. Planting of selected plant at selected place is done at an early stage, stopping of branches is done for dense growth and then frame is placed. Maintenance of topiary plant need special care plant should be liberally watered and fertilized so that plants make a dense and colourful growth.

Chapter 7

Other Features of a Garden

Garden Walls

A garden lover will never like to block the view of his garden by putting a wall along its periphery. A garden as it is enjoyed from inside should also be visible from outside.

- ☆ Low brick or concrete or stone wall of say 60-90 cm height and to put over it some grills.
- ☆ Alternatively walls from 1.8-5.0 m may also be erected depending upon the size of garden.
- ☆ One may grow creepers such as *Ficus repens* over the wall.

Flower Beds

The most essential criterion of a flower beds is display the flowers in the best way possible. It may be born in mind that flowers look best when massed in a bed. The flower beds are of special importance under condition of the plains of India since they can be kept planted through out the year unlike those in England or other parts of Europe where nothing can be grown in the open between November-March. Annuals and herbaceous, perennial flowers, flowering plants are grown in flower beds to provide massing effect of different colours borders is continuous beds of more length than width containing plant of one kind only.

- ☆ The flower beds display the flower in their best way.
- ☆ More important part of a formal garden.
- ☆ These should be simple in design as circular, rectangle, and square.
- ☆ It should be simple in design as it is easier to maintain them.

- ☆ It is most important part of a formal garden.
- ☆ It can be planted during September-October (winter flower), February-March (summer flower) and April-June (rainy flower).

Path or Garden Drives

A garden must have a carriage drive leading to the house and the garage besides several other paths or walks leading to different parts of the garden. When the planting of low growing perennial plants are grown on the paths it is the also become a part of formal and in formal landscape design. The garden path should never be less than 60 cm wide but should be preferably between 90-120 cm, if sufficient is a available. The ideal plant should be quick growing with small stands against trained which is done frequently. A garden drives leads to the house and the garage while paths walk leads to different part of garden. The principle of construction of garden paths are the same as for drives but since these are narrow and not meant for heavy traffic the foundation need not be so deep.

- ☆ Paths should not be generally too high or too low from the adjoining ground except in a marsh or a rock garden.
- ☆ Garden paths are generally made of gravel, paving stone, crazy paving, bricks, grass, *etc.*
- ☆ Garden drives are generally made of gravel and asphalt or concrete.

Steps

Sometimes it becomes necessary to have steps in a garden as for example when a path goes from one level to the other or one has to climb or get down from the terrace garden. The steps in the garden should be different than those in the building.

- ☆ Steps used to move from one level to others.
- ☆ To climb or get down from the terrace garden
- ☆ Steps can be made of concrete, stones, wood or gravels.
- ☆ It should be quite broad and raises low.

Arches

A garden may need some arches for training climbers or ramblers. A general rule is to have the same width as the height provided space permits.

- ☆ Arches are generally erected over walk usually at the entrance and are usually 2 m in height.
- ☆ Generally constructed near the gate all over the paths.
- ☆ The supporting poles can be made of stone pillar.
- ☆ It should be at least 2-2.5 m height.

Pergolas

For growing creepers in a row pergolas are ideal structures on which these may be trained. A pergola may be defined as a series of arches joined together.

- ✰ It is series of arches joined together generally constructed over pathways.
- ✰ It adds beauty to the garden.
- ✰ It is useful as resting place.
- ✰ In broad pergola it may also be possible to keep a few shade-loving plants to protect them against sun but this may not be desirable all the time.
- ✰ Made of wood or stone or brick pillars, angled iron and galvanized pipe.

Terraces

In hilly tract it is not possible to have a large piece of land in one plane for laying a garden and hence gardens are laid in terraces, where it is a natural phenomenon. But in plain place the land for gardening may not have any natural undulation for terracing.

- ✰ This is common feature in the English or Japanese garden.
- ✰ It is the raised flats are gravel section.
- ✰ To construct a part or a major portion of the garden.
- ✰ It forms a natural links.
- ✰ It should give a full view of the garden.

Paved Garden

A paved garden, if properly laid, can be a very attractive feature of a garden. There are some specific plants, which adopt themselves well to a paved garden. These should be dwarf in nature and stand a considerable amount of water and tear from shoes of different weight. But a paved garden should be laid in a path which is not used very often. A special paved garden may also be created if a path suitable for this is not available. Ordinarily a paved garden is meant for walking, although not very frequently and hence the interstices of the paved garden should be planted sparingly.

Carpet Bedding

The art of growing ground cover plants closely and trimming them to a design or alphabetical latter is called carpet beds. Colourful foliage as hedge plants is found to be more suitable for this purpose, *e.g.,* Alternanthera, Cinararia, Coleus, Irisine, Portulaca, *etc.*

- ✰ It is used to cover an area preferably a bed or a series of beds.
- ✰ It is arranged in a slope or a slanting position.

- ☆ It is the form of a figure or some letters that's cut out.
- ☆ Plants having different growth habits or having different colourful leaves are used.

Dry Wall

The term dry wall is rather a misnomer as a wall may be dry but once the plants are planted over it, it no more remains dry but becomes an object of beauty. Whatever may be the name, garden laid on a wall or wall planted with different plants are termed as dry wall in the garden dictionary.

Plants growing in the crevices of the stones and hanging down the face of a dry wall look beautiful and are becoming a common feature of the English garden.

Chapter 8

Annuals

Annuals or seasonal are the group of plants which complete their life cycle in one season or in one year. Annuals are commonly known as seasonal as they complete the process of life like germination, growth, flowering and seed formation in one season and finally the plant wither out. Annuals are widely used for garden decoration, cut flower and pot plants. They are frequently grown as bedding plant and in rock garden on the sides of lily pool and in shrubberies.

Important Feature of Annuals

- These plants can be easily grown.
- A wide variation exist among plants in form, habit of growth, flower colour and flower size.
- Propagation through both sexual and asexual.
- Annuals without aid of other plants keep the garden full and joyful year round.
- Annuals are grown in pots or in ground.

Classification of Annuals

(1) According to Season

Annuals have different temperature requirement for their growth, development and flowering according:

- Winter
- Summer
- Rainy

a. Winter Season Annuals

The annuals that can be easily grown under low temperature with profuse flowering falls under winter annuals. The nursery should be prepared during month of September. Whereas, transplanting should be done in October in hilly area planting during the month of March, *e.g.,* Calendula, Ice plant, Pansy, Phlox, Portulaca, Sweet sultan, *etc.*

b. Summer Season Annuals

The annuals that thrive well under extremely high temperature during summer fall under summer annuals. Sowing should be done during the end of February and beginning of March and transplanting during the month of March to April, *e.g.,* Cosmos, Gaillardia, Kochia, Sunflower, Zinnia, *etc.*

c. Rainy Season Annuals

The annuals that can withstand rains and high humidity come under rainy humidity. Sowing time is June and transplanting in July, *e.g.,* Amaranthus, Balsam, Gaillardia, Cocks comb, Zinnia, *etc.*

(2) According to Height

In general medium and dwarf annuals are ideal for growing in pots. Whereas, tall annuals can be used for fence. Depending on their height they can be classified as follows:

a. Dwarf

Annuals that can grow up to height of 0.45m comes under dwarf category. For edging, these annuals are used, *e.g.,* Calendula, Pansy, Sweet alyssum, French marigold, Kochia, *etc.*

b. Medium

The annuals whose height varies between 0.45-0.75 m is medium and used for screening purpose, *e.g.,* Petunia, Salvia, Candy tuft, Aster, Carnation, *etc.*

c. Tall

These annuals are of height more than 0.75m. These are commonly used for screening purpose, *e.g.,* French marigold, Cosmos, Holly hock, Sun flower, *etc.*

(3) According to Colour

Annuals have got wide range of flower colour in pure form–a single or in combination.

a. Blue and Purple Colour

Sweet pea, China aster, Ageratum, Pansy, Corn flower, *etc.*

b. White Colour

Candy tuft, Petunia, Phlox, Sweet sultan, *etc.*

c. Pink and Red Colour

Salvia, Phlox, Holly hock, Verbera, *etc.*

d. Yellow Colour

Calendula, Sun flower, Tithonia, Cosmos, Marigold, *etc.*

(4) According to Commercial Value

a. Cut Flower

Cut blooms of seasonal are used as vase decoration, *e.g.,* Aster, Calendula, Carnation, Sweet pea, Zinnia, *etc.*

b. Loose Flower

Some annuals are used as loose flower for making garland and veni, *e.g.,* Annuals chrysanthemum, Marigold, Amaranthus, Gaillardia, Balsam, *etc.*

(5) According to Situation

Different annuals are suited to different situation and therefore right type of annuals should be selection for particular purpose.

a. **For carpet bedding-** Candytuft, Ice plant, Salvia, Sweet alyssum, Sweet william, *etc.*
b. **For climbing-** Sweet Pea, Clitoria ternatea, Nasturtium, *etc.*
c. **For dry flower arrangement-** Nigella, Statis, Acrolium, Cocks comb, *etc.*
d. **For edging-** French marigold, Pansy, Sweet Alyssum, Ageratum, Candy tuft, *etc.*
e. **For fragrance-** Sweet sultan, Sweet william, Sweet alyssum, Stock, Jasmine, *etc.*
f. **For hanging basket-** Petunia, Sweet alyssum, Toroia verbena, *etc.*
g. **For ornamental foliage-** Amaranthus, Kochia, *etc.*
h. **For pinching requirement-** Marigold, Ageratum, Zinnia, Petunia, Calendula, *etc.*
i. **For pots-** Balsam, Kochia, Amaranthus, Petunia, Portulaca, *etc.*
j. **For rockery purpose-** Candy tuft, Ice plant, Nasturtium, Phlox, Portulaca, Pansy, *etc.*
k. **For screening purpose-** Nasturtium, Holly hock, Sweet pea, Morning glory, *etc.*
l. **For shady situation-** Phlox, Salvia, Balsam, Torania, Ageratum, Cineraria, *etc.*
m. **For window purpose-** Aster, Calendula, Petunia, Phlox, *etc.*

(6) According to Soil Requirement

a. **For very poor soil-** Cocks comb, Gaillardia, Balsam, Portulaca, Petunia, Sweet sultan, Sweet alyssum, *etc.*

b. **For alkaline soil-** Balsam, Phlox, Nasturtium, Zinnia, Poppy, *etc.*

c. **For acidic soil-** French marigold, Zinnia, Nasturtium, *etc.*

Cultivation Practice

Soil

Annuals grow well in well drained fertile sandy loam soil rich in organic matter. Soil should not be acidic or alkaline. The ideal pH 6-7.5 for best cultivation of annuals. In case of clay and sandy soil addition of organic manure will be helpful to improve the soil texture and porosity.

Nursery Management

All the annuals are propagated by seeds. Seeds can be sown in nursery beds, earthen pot, seed pan, seed tray, *etc.*

Management

- ☆ Seed bed should be propagated in a place with good drainage system and away from shade.
- ☆ Soil should be sandy loam and natural in pH.
- ☆ FYM at 10 kg/m^2 should be incorporated in bed.
- ☆ Size of seed bed should be 3x1 m^2 bed length increase with as per requirement.
- ☆ In seed bed are sown in rows 6 cm apart.
- ☆ After sowing seed must be covered with fine FYM and paddy straw.
- ☆ Watering should be done soon after sowing and latter when required.
- ☆ In general seedling will be ready to transplant after one month when attend 4-5 true leaf.

Annuals Bed and Transplanting

Time of transplanting must be during evening hours since the night temperature is beneficial for the successful establishment for seedling. Before transplanting seedling must be harden by check the irrigation for few days.

Description of some Important Flowering Annuals

Sl.No.	Common Name	Botanical Name	Family	Flower Colour	Plant Height (cm)	Remark
A.	**Winter season annuals**					
1	Paper flower	*Acroclinum roseum*	Compositae	White and pink	45-60	Daisy-like- flower
2	Floss flower	*Ageratum mexicanum*	Compositae	Blue, white pink	20-45	Puff hands of flowers
3	Hollyhock	*Althaea rosea*	Malvaceae	Pink, scarlet red, matte violet and yellow	100-120	Majestic plant
4	Sweet alyssum	*Alyssum maritimum*	Cruciferae	White, pink lilac	10-20	Edging and bedding purpose
5	Snapdragon	*Antirrhinum majus*	Scrophulariaceae	White, pink, yellow frock red and maroon	30-70	Flowers like dragons jaw on spikes
6	Daisy	*Bellis perennis*	Compositae	White, pink and crimson	20-30	Dwarf plants
7	Swan river daisy	*Brachycome iberidifolia*	Compositae	White, pink	20-50	Dwarf plant
8	Calendula	*Calendula officinalis*	Compositae	Yellow orange	30-50	Meaning first day of the month
9	Corn flower	*Centaurea cyanus*	Compositae	Blue, pink white	60-80	Weed of the corn field
10	Wall flower	*Cheriranthus cheiri*	Cruciferae	Yellow, burnt orange	30-45	Bedding and pot culture
11	Annual chrysan-themum	*Chrysanthemum coronarium*	Compositae	Yellow, whit	90-120	Bedding purpose
12	Cineraria	*Senecio cruentus*	Compositae	Purple, white	30-45	Pot culture, shade loving
13	Clarkia	*Clarkia elegans*	Onagraceae	White, pink, rose scarlet	60-80	Have long spikes
14	Parrot bill	*Clanthus dampieri*	Leguminosae	Crimson	60-80	Flower are pendulous
15	Coreopsis	*Coreopsis tinctoria*	Compositae	Yellow crimson, brown	45-60	For bedding purposes
16	Cosmos	*Cosmos bipinnatus*	Compositae	Pink, purple, crimson, white	60-100	Good for mass planting
17	Dahlia	*Dahlia variabilis*	Compositae	Yellow, red, blue, white, crismon	60-120	Bedding and pot culture
18	Larkspur	*Delphinium ajacis*	Ranunculaceae	Violet, crimson, pink, blue, white	45-20	Bedding and cut flower

Sl.No.	Common Name	Botanical Name	Family	Flower Colour	Plant Height (cm)	Remark
19	Sweet william	*Dianthus barbtus*	Caryophyllaceae	Pink, crimson, mauve, yellow	30-45	Scented flower
20	Carnation	*Dianthus caryophyllus*	Caryophyllaceae	Pink,white, yellow, mauve, crimson violet	30-60	Cut flower good vaselife
21	African daisy	*Dimorphotheca calendulacea*	Compositae	Yellow, white	30-60	Flowers open during day time
22	California poppy	*Eshscholzia californica*	Papaveraceae	Yellow, orange	30-45	Bedding and pot culture
23	Blanket flower	*Gaillardia pulchella*	Compositae	Yellow, orange, lemon, maroon	30-45	Bedding and cutting
24	Treasure flower	*Gazania splendens*	Compositae	Yellow, orange white,red	20-25	Rock garden
25	Boys breath	*Gypsophila elegans*	Caryophyllaceae	White, pink	30-45	Small lance shaped flower
26	Everlasting flower	*Helichrysum bracteatum*	Compositae	Yellow, pink, red	45-60	Flower long lasting
27	Candytuft	*Iberis amara*	Cruciferae	White, lilac	20-30	Flower appear in tufts on spikes
28	Sweep pea	*Lathyrus odoratins*	Leguminosae	White, blue, pink, mauve	90-120	Grown as background in tufts on spikes
29	Statice	*Limonium sinuatum*	Plumbaginaceae	White, pink, purple	45-60	Good cut flower
30	Linaria	*Linaria bipartite*	Scrophulariaceae	White, blue, pink, red yellow	30-45	Bed and pot culture
31	Lupine	*Lupines lurens*	Leguminosae	Yellow, blue	40-60	Cutting beds borders
32	Stock	*Mattihola* sps.	Cruciferae	White, red	60-90	Scented cutting
33	Ice plant	*Mesembryanthemum tricolor*	Aizoaceae	Pink, white, yellow	20-30	Rock garden dry wall
34	Nemesia	*Nemesia strumosa*	Scrophulariaceae	White, blue, yellow, orange	45-60	Cut flower
35	Nigella	*Nigella demascena*	Ranunculaceae	White, blue, rose	45-60	Good colour for cut flower
36	Shirley poppy	*Papaver roheas*	Papaveraceae	Scarlet pink white	60-75	Popular flower
37	Phlox	*Rhloxy drummondii*	Polemoniaceae	Pink, crimson, purple, violet	30-45	Bedding plant

Sl.No.	Common Name	Botanical Name	Family	Flower Colour	Plant Height (cm)	Remark
38	Nasturtium	*Tropacolum majus*	Tropacolaceae	Yellow, orange, red	30-40	Bedding
39	Salvia	*Sativa splenders*	Labiatae	Red, white, blue, scarlet	30-45	Shade loving
40	Saponaria	*Saponaria calabrica*	Caryophyllaceae	Pink, white, red	60-90	Flowers resemble gypsophila
41	Pansy	*Viola tricolor*	Violaceae	Purple, blue, yellow	20-30	Pot culture
42	Aster	*Callistephus chinensis*	Compositae	Pink, rose, purple	30-45	Pot and bed culture
43	Sweet sultan	*Centarrea moschata*	Compositae	Purple, pink, blue	30-45	Bed culture
44	African marigold	*Tagetes erecta*	Compositae	Yellow, orange	45-100	Bed culture
45	French marigold	*Tagetes patula*	Compositae	Red, brown, yellow, orange	20-30	Pot and bed culture
46	Verbena	*Verbena hybrid*	Varbinaceae	Pink, red, purple, blue	30-45	Bed culture
47	Lace flower	*Trachymene caerule*	Umbelliferae	White. Pink, blue	45-60	Umbrella like clusters
48	Linum	*Linum grandiflorum*	Linaceae	Red, purple	30-45	Bed culture
B.	**Summer season annuals**					
1	Portulaca	*Portulaca grandiflora*	Portulaceae	Pink, red, yellow, blue, orange	10-15	Pot and bed
2	Blanket flower	*Gaillardia punechella*	Compositae	Yellow, brown, orange, scarlet	45-60	Easy to grow
3	Kochia	*Kochia scoparia*	Chenopodiaceae		60-75	Green leaves
4	Petunia	*Petunia hybrida*	Solanaceae	Pink, blue, purple, white	20-30	Give good effect
5	Zinnia	*Zinnia elegans*	Compositae	Violet, orange, white	70-80	Very hardy, easily grown
6	Sada bahar	*Catharanthus roseus*	Apocynaceae	Pink	30-60	Very hardy plant
7	Coreopsis	*Coreopsis tinctoria*	Compositae	Yellow, scarlet	45-60	Very attractive flower
8	Sunflower	*Helianthus annuus*	Compositae	Yellow, orange	60-120	Can be grown throughout the year
9	Cosmos	*Cosmos bipinnatus*	Compositae	Pink, orange, yellow	60-70	Bedding purpose

Sl.No.	Common Name	Botanical Name	Family	Flower Colour	Plant Height (cm)	Remark
C.	**Rainy season annuals**					
1	Balsam	*Impatiens balsamina*	Balsaminaceae	Pink, red, rose	60-70	Very delicate
2	Cocks comb	*Celosia argentea* var. *cristata*	Amaranthaceae	Red, yellow orange	30-60	Hardy plant
3	Zinnia	*Zinnia elegans*	Compositae	Violet, orange	70-80	Hardy plant
4	Amaranthus	*Amaranthus caudatus*	Amaranthaceae	Pink, white	45-60	Grown in pots
5	Corn flower	*Centaurea cyanus*	Asteraceae	Blue, Purple	40-50	Grown as ornamental plant

Planting Distance

Distance of annuals will be depend upon the height of the plant. If distance is not properly maintained the growth of the plant will be adversely affected. The distance for dwarf annuals is 30×30cm, medium 45×45cm and tall 60×60cm should be maintained.

Manures and Fertilizers

Application of FYM at 5 kg/m^2 should be done during bed preparation and fertilizer NPK (15:10:10gm/m^2) should be applied after 30 day of transplanting. After transplanting of seedling bed must be weed free and irrigate properly.

Intercultural Operation

In annuals 3-4 weeding can be done and hoeing and earthing to be followed if necessary.

Plant Protection Measure

Disease- Damping off, leaf spot, leaf blight, powdery mildew, downy mildew and wilt.

Control- Capton 2gm/lt, Carbendazim 2 gm/l, Dithen-M-45, Dithene-Z-78 2 gm/l.

Insect- Aphids, Bettles, Weevils, and Catter piller.

Control- Imidacloparid 1.5-2.0 ml/l.

Chapter 9

Annuals Border

In this type of border annuals are planted according to their height to beautify the garden. Annuals border are found two types:

- ✰ Single face border
- ✰ Double face border

Single Face Border

In this type dwarf plants is kept on front of beds, medium height annuals are planted in the centre while the taller annuals are planted in the back.

Double Face Border

In this type tall annuals are planted in the center and medium and dwarf annuals are planted in both sides of tall annuals so that full view can be seen from both sides.

Herbaceous Border

In this plants are arranged in irregular group of a kind for harmonious contrasting colour effect, either all flowering at one time or successfully in such a way that those flowering latter should grow up and screen the all ready flowered. The planting of annuals in the border of a plot is called as herbaceous border. When the border is to be viewed from both side tall plant are planted in center followed by medium and dwarf planting in both sides. Herbaceous border may include all the herbaceous plant including herbaceous perennial (Flowering and non-flowering), bulbous and annuals.

Plant Suitable for Herbaceous Border

A. According to Height

1. **Tall-** Corn flower, Holly hock, Sun flower, Dahlia, *etc.*
2. **Medium-** Aster, Salvia, Geranium, Gladiolus, Lillium, *etc.*
3. **Dwarf-** Phlox, Iris, Ice plant, Kochia, Gerbera, Anthurium, *etc.*

B. According to Colour Scheme

1. **Monochromatic colour-** It consists of different tempts and shade of one colour and is seldom achieved in its pure from in the landscape. Colour scheme could include white and pink flower with back ground of dark pink and red brick house.
2. **Analogous colour-** It combines colours which are adjacent side by side on colour wheel. An analogous colour scheme includes green, blue green, green blue, blue violet blue.
3. **Complementary colour-** Red and green would be complementary colour. A complementary colour scheme may be achieved by using plants with green foliage against a red brick house. Transition is gradual change in colour can be illustrated by redial sequence on the colour wheel (Monochromatic colour scheme).

Shrubs

Shrubs is a woody plant with several stem and branches arising at the ground level from the main stem. This is perennial in nature and smaller than a tree. Shrubs are very important in the garden as flower shrubs produce beautiful flower at eye level and fragrant shrubs emit fragrant as nose level. Shrubs produce flower, foliage, fruits and berries and enhance beauty of garden. They are hardy in nature and require less care than other ornamental plants.

Importance of Shrubs

- ☆ Enhance beauty of surrounding.
- ☆ Provide fragrance in garden.
- ☆ They take the place of garden boundary wall and provide live lines to the garden.
- ☆ Divides different areas are feature in the garden.
- ☆ Screen off unwanted sides.
- ☆ Reduces wind velocity, soil erosion, weed growth.

Classification of Shrubs

1. According to Beauty of Plants

(a) **For flowers-** There are several shrubs that are grown for their very attractive flowers and enhanced beauty of garden, *e.g., Barleria cristata, Bougainvillea* species, *Crossandra undulaefolia, Hibiscus rosa-sinensis, Jasminum pubescence, etc.*

(b) **For foliage-** These shrubs produce beautiful foliage, *e.g., Codiaem varigatum, Euphorbia cotinifolia, Manihot species, Aralia, etc.*

(c) **For variegated foliage-** There are certain shrubs that produce variegated foliage, *e.g., Duranta plumier, D. varigata, Manihot utilissima* var. *Tebernaemontana coronaria, etc.*

(d) **For flower and foliage-** In these classes are fund as *Acalypha hispida,* Bougainvillea (Archana, Bhabha), *Hibiscus* (Snow queen and Redhot), *Hamelia patens, etc.*

(e) **For fruits-** These are a shrub that bears showy fruits, *e.g., Carissa carendus* (Dark red), *Citrus japonica* (Yellow), *Duranta macrophylla* (Yellow), *Rauwolfia canscess* (Red), *etc.*

(f) **For fragrance-** In which the fragrance flowers are grown for beautification, *e.g., Cestrum nocturnum* (Rat-Ki-Rani), *Cestrum docturnum* (Din-Ka-Raja), *Jasminum auriculatum, Jasminum grandiflorum, Jasminum sambac, etc.*

2. According to Requirement of Sunlight

(a) **For sunny situation-** *Bougainvillea, Hibiscus rosa-sinensis, Jasminum sambac, Jasminum grandiflorum, Lanata camara, Murraya exotica, etc.*

(b) **For partial or semi shade-** *Jatropha rosia, Mussaend philipica, Maglolia grandiflora, Nandina domestica, etc.*

(c) **For both partial and sunny shade-** Acalypha, *Cesturm nocturnum, Chrosandra* sps., *Hemelia patens, Thunvergia erecta, etc.*

3. According to pH

(a) **Shrub tolerant to acidic soil-** pH 5.0 below, *e.g., Juniperus communis, Rhododendran* sps. *Kalmia latifolia, Aesculus parviflora, etc.*

(b) **Medium acidic soil-** pH 5.0- 6.0 above, *e.g., Gardenia jasminoides, Hamamelis virginiana, Itea virginica, Hydrangea macrophylla, etc.*

(c) **Shrub tolerant to alkaline soil-** pH between 6.5 -7.5, *e.g., Berberis* sps., *syringe* sps., *Hibiscus* sps., *Viburnum* sps., *etc.*

4. According to Height

(a) **Dwarf-** Dwarf type shrubs are found up to 1 m. *Barleria cristata* (violet blue), *Crossandura undulaefolia* (Yellow), *Erathemum laxiflorum* (purple-rose), *Jasminum sambac* (white), *Lantana camera* (blue), *etc.*

(b) **Medium-** The hight of plants between 1.0 to 2.5m. *Acalapha hispida* (Red), *Allamanda neriifolia* (Yellow), *Cestrum nocturnum,* (White), *Cestrum docturnum* (White), *Jasminum multiflorum* (White), *Lantana camara* (White), *etc.*

(c) **Tall-** The hight of plants found tall between 2.5 to 4m. *Buddelia asiatica* (white), *Clerodendrum inerme* (white), *Gardenia jasminoides* (white), *Hamelia patens* (Red), *Hibiscus rosa sinensis* (Red), *etc.*

Purpose of Growing

Specimen

Exceptionally beautiful shrub can be planted as single specimen in a lawn, area near their special quality or put on the show, *e.g., Bougainvillea* sps. *Camelia japonica, Hibiscus rosa sinensis, Hameia patens, etc.*

Standard and Half-standard

Several shrubs may be trained and allowed to develop a single branch at a center height. Single stem 1meter for standard 0.5meter for half standard and single stem plant should be selected for making standard side branch should be removing. Standard and half-standard plant look more attractive when planted along the path and drives in a formal garden.

Shrubbery

An area of cultivated shrubs in a park or garden is known as shrubbery or any area in the garden fully solely devoted to the shrub plantation is called shrubbery. These shrubs are maintained in formal way therefore require regular training, pruning and clipping. Maintenance of shrubbery is easy and become less costly if planted properly considering following point:

- ✰ Selection of hardy shrubs.
- ✰ Shrubs having profuse flowering and beautiful foliage.
- ✰ Selection of shrub according to soil.
- ✰ Shrub having slow growing habit and required less pruning and management.

Principles and Planning of Shrubbery Border

- ✰ Shrubs should be planted in east, south or south-west direction to achieve best result.
- ✰ In a large garden shrubbery should be arranged in more area to reduce space and maintenance cost of garden.
- ✰ Shrubbery placed in the front of tall tree or along the conifers looks very appearing.
- ✰ Flower and foliage colour should be visible from a distant place when tall, medium and dwarf shrub are arranged in the shrubbery.

☆ Along the path, drives, terrace lawn and in front of house small shrub should be planted. Whereas, in front of big tree taller shrubs should be planted for picturesque effect.

Arrangement of Shrubbery

Shrubbery is arranged basically in two way, *i.e.,*

☆ According to height

☆ According to colour

(1) According to Height

In front of an object, *i.e.,* tree, big building. Taller shrubs are planted first, then medium shrubs to be planted after that smaller shrubs are placed in the shrubbery. Shrubs are arranged in two ways as give below:

(a) **Single face shrubbery border-** In this case plantation of shrubs is done in front of trees on boundary. Shrubs are arranged according to height then the taller shrubs in the last then medium and after that smaller shrubs should be planted.

(b) **Double face shrubbery border-** Its observed from both sides in this case taller shrubs are planted between the two medium shrubs besides with smaller shrubs are planted in both sides.

(2) According to Colour

These shrubs have attractive colour light yellow, white, golden yellow, pink, scarlet, crimson, rose red, violet, blue, *etc.*

Three common colour schemes are given below:

(a) **Monochromatic colour scheme-** Massing of single or one colour is called monochromatic colour scheme, *e.g.,* White- *Jasminum sambac* (dwarf), *Cestrum nocturnum* (medium), *Gardenia josminoids* (tall) and Yellow- *Galphimia gracilis* (dwarf), *Allamanda nerifolia* (medium), *Thevetia peruviana* (tall).

(b) **Analogous colour scheme-** It is also called harmonious colour scheme in this scheme shrubs are planted with closely related colour like white, pink, red colour flower or vice versa.

(c) **Contrast colour scheme-** It is also called complementary colour scheme. Two opposite colour shrubs are planted in this scheme like blue-orange, red-green and yellow-violet.

Rock Garden

Rock garden is made by the rock or stone, garden look very natural, no symmetrical and unleveled all the garden elements and adornments can successfully used garden with any restriction. It is completely built of industrial and home waste and through away items. Rock garden of Chandigarh is also known as Nek Chandra rock garden after its founder Nek Chandra. A government official started it in 1957 in area 40-50 acres. Plants are- *Jatropha podagrica, Lantana depressa* and *Russellia juncea*

Edging- *Samchezia nobilis* and *Lantana sellowiana*

Hedges- *Durenta plumieri, Murraya panniculata* and *Hamelia patens*

Screening- *Jasminum pubescence* and *Duranta repens*

Ground cover- *Lanatana sellowiana* and *Smchezia nobiliss*

Topiary- *Clerodendron inermi, Cupressus macrocarpa* and *Duranta plumier*

Pot plants- These shrubs are placed in verandah, terraces, balcony or other place. Size of pot should be of 12-15cm, potting medium consist of soil manure, leaf mould FYM 2:1:1 (S:M:L), *e.g., Citrus japonica*, Croton, Poinsettia, *etc.*

Planting Care and Management of Shrubs

Propagation

1. **By seed-** Seeds are collected when they are fully ripe. They are dried in the shade and stored in airtight bottle in dry places. In rainy season nursery are prepared and seed are sown, *e.g., Colliandra species, Thevetia peruviana* and *Jatropha species.*
2. **By cutting-** The best season for propagation by cutting in rainy season although cutting can be made in Feb-March if water supply is not limited. The cuttings are made at least 3-5 nodes and about 15-20cm length, *e.g., Hibiscus rosa-sinensis, Jasminum sambnac, Bougainvillia* species, *Cestrum nocturnum.*
3. **By layering or air layering-** In rainy season layering done by removing the bark of 0.5-3.5 cm long should be remove and stem or branches of shrubs are 20-30 cm long should be selected and wrapping garden soil and sand mixture, *e.g., Jasminum sambac, Bougainvillea*, Croton and *Ixora* species.

Soil and Site Preparation

Shrubs are hardy in nature therefore it can be grown in less fertile and neutral pH soil. Before planting the shrubs soil should be well prepared by addition of organic manure @ 5-6kg/m^2 area.

Planting

The ideal time for planting is rainy season but if irrigation facility is available it can be planted in February-March. In shrubbery border each type of shrubs should be planted at list 2-3 rows in respective distance. Normally the tall shrubs are planted at 1.5-2m distance, medium shrubs are planted 1-1.5m distance and the dwarf shrubs are planted 0.5-1m distance.

Description of some Important Flowering Shrubs

Sl. No.	*Common Name*	*Botanical Name*	*Family*	*Flower Colour*	*Flowering Time*	*Method of Propagation*	*Remark*
1	Gandharaj	*Gardenia jasminoides* syn. *G. florida*	Rubaceae	White	April-June	Air-layering cutting	Popular background & fragrant 3m tall
2	Gurhal	*Hibiscus rosa sinensis*	Malvaceae	Rose-scarlet		Air-layering cutting	Bushy, 1-2m tall
3	Thal kamal	*Hibiscus mutabilis*	Malvaceae	White in the early morning and changes to pink and deep rose as the day advances	Throughout the year	Seeds cutting	3m tall shrub
4	Rukmini/Pak Tak	*Ixora coccinea* *I. singaporensis*	Rubiaceae	White scarlet Scarlet	April-June/July-Sept, Through the year, Greater parts of the year	Cutting/Layering	2m tall bushy
5	Mussaenda	*Mussaenda frondosa* *M. luteola* *M. philippica*	Rubiaceae	White Yellow Pink	April-Sept April-Sept April-Sept	Layering Layering Layering	Beautiful shrub Beautiful shrub Beautiful shrub
6	Kamini	*Murraya exotica*	Rubiaceae	White	Sept-Oct	Seed/Layering	Used for hedge topiary
7	Karonda	*Carissa carandas*	Apocynaceae	White		Seed	Used for hedge, berries
8	Krishnachura	*Caesalpinia pulcherrima*	Leguminosae	Orange-scarlet	Throughout the year	Seed	Beautiful shrub
9	Radhandra	*Caesalpinia pulcherrima var. flava*	Leguminosae	Yellow	Appril-Aug	Seed	Beautiful shrub
10	Calliandra	*Calliandra inaequilatera*	Leguminosae	Red, pink white		Seed/layers	Puff like flower
11	Bougainvillea	*Bougainvillea* spp. *glabra*	Nycataginaceae	Various colour	Feb-June/Sept-Dec	Cutting	Versatile shrub

Sl. No.	Common Name	Botanical Name	Family	Flower Colour	Flowering Time	Method of Propagation	Remark
12	Allamanda	*Allamanda cathartica*	Apocyraceae	Yellow	April-Aug	Cutting-layering	Can be used as climber
13	Clerodendron	*Clerodendron inerme*	Verbenaceae	White	July-Aug	Cutting	Used for hedge
14	Raat-ki-rani	*Cestrum nocturnum*	Solanaceae	Creamy yellow	April-July	Seed/cutting	Fragrant during night bushy, quick growing
15	Day- Queen	*Cestrum diurnum*	Solanaceae	White	Summer	Seed/cutting	Bloom during day time
16	Crossandra	*Crossandra* spp.	Acanthaceae	Yellow orange, brick red	Feb-June	Seed/cutting	Flower are used for making gajras
17	Golden dew drop balchari	*Duranta plumieri*	Verbenaceae	Blue		Seed/cutting	Used for making hedge
18	Gandharaj	*Gardenia jasminoides* syn. *E. florida*	Rubiaceae	White	Feb-July	Cutting, layering	Have fragrant flower
19	Hamelia	*Hamelia patens*	Rubiaceae	Orange red	April-Aug	Cutting, layering	Used for making good hedge
20	Barbados cherry	*Malpighia glabra*	Malpighiaceae	Purple	April-Aug	Cutting seed	Specimen plant, cherry like fruit
21	Kaner	*Nerium oleander*	Apocynaceae	White, pink, red	April	Cutting layering	Popular shrub
22	Chitra	*Plumbago caoensis* syn. *P. auriculata*	Plumbaginaceae	Blue	Feb-Sep	Cutting layering	Used for edging
23	Poinsettia	*Poinsettia pulcherrima*	Euphorbiaceae	Red treats	Dec- Feb	Cutting	Rapid growing sun loving
24	Tulsi	*Rauwolfia chinensis*	Apocynaceae	White	Dec-Feb	Seeds	Evergreen medicinal shrub
25	Coral pant	*Russelia juncea*	Scrophulariaceae	Coral red	March-Aug	Cutting	Pendulous branches
26	Chandni, Tagar	*Tabemaemontana coronaria*	Apocynaceae	White	March-Aug	Cutting	Used for making hedge
27	Pila kaner	*Thevetia nerifolia*	Apocynaceae	Yellow	Throughout the year	Seeds/cutting	Funnel-shaped flower
28	Bela	*Jasminum sambac*	Oleaceae	White	May- June	Cutting	

Description of some Important Foliage Shrubs

Sl.No.	*Common name*	*Botanical name*	*Family*	*Method of propagation*	*Remark*
1	Aglaonema	*Anglaonema* spp.	Araceae	Cutting division	Perennial, leaves green with marking of grey or variegated or silver.
2	Alocasia	*Alocasi* spp.	Araceae	Cutting tubers, rhizomes	Beautiful, hardy, indoor plant
3	Aralia	*Aralia* spp.	Araliaceae	Cutting	Hardy plant
4	Sataber	*Asparagus* spp.	Liliaceae	Seed suckers	Beautiful bristle like cladodes
5	Colocasia	*Colocasia* spp.	Araceae	Tubers	Attractive foliage
6	Cordyline	*Cordyline* spp.	Liliaceae	Node cuttings suckers	Related to Dracaena
7	Dracaena	*Dracaena* spp.	Liliaceae	Node cuttings suckers	Long leaves
8	Cycas	*Cycas* spp.	Cycadaceae	Suckers	Good for pod, greenhouse and indoor plant
9	Dieffenbachia	*Dieffenbachia* spp.	Araceae	Cutting	Pot plant, warm, humid conditions plant are poisonous
10	Rubber plant	*Ficus elastica*	Moraceae	Air layering, cutting	Pot plants
11	Maranta (calathea)	*Maranta* spp.	Marantaceae	Suckers, cutting	Good pot plant
12	Monster	*Monster acuminate*	Araceae	Stem cutting	Root climbers with good foliage
13	Philodendron	*Philodendron* spp.	Araceae	Stem cutting	Different shapes of foliage
14	Pilea	*Pilea muscosa* syn.*o. microphylla*	Urticaceae	Stem cutting	Grown in shade place
15	Sansevieria	*Sansevieria cylindrica* *Sansevieria trifasciate*	Liliaceae	Division for suckers	Perennial, erect, sword like leaves
16	Tradescantia	*Tradescantia albiflora*	Commelinaceae	Stem cutting	Low growing, trailing or creeping

Irrigation

First irrigation should be given just after planting. Frequent irrigation should be given till the date of establishment. In early age of planting frequent irrigation will be needed. During winter season irrigation should be given at monthly interval whereas, in summer season the plant should irrigate 15 days interval. In rainy season no irrigation is required.

Gap Filling

In gap filling the replanting of the plant will be necessary for filling the gap due to mortality.

Weeding

Weeding in shrub planting field is essential after two years of establishment.

Pruning

For providing proper shape and enhance flowering pruning is essential. Pruning should be done in the month of December-February in the shrubs like hibiscus, jasmine and hydrangea. Pruning should be done after flowering.

Chapter 10

Trees

A tree is a perennial plant with an elongated stem or trunk, sprouting branches and leaf in most species. Trees trend to be long leaved living up to several 100 year.

TREE- logical meaning of each latter

T – Temperature and micro-climate moderation.

R – Removal of air pollutant.

E- Erosion control.

E – Energy conservation.

Trees are selected for landscaping for following factors:

- ☆ Habitat
- ☆ Habit
- ☆ Form
- ☆ Height
- ☆ Spread
- ☆ Trunk height
- ☆ Texture (Fine, Medium)
- ☆ Growth rate (Slow, Medium and Fast)
- ☆ Flower colour

General Importance

1. Trees provide additional necessities such shelter, medicine and tools, *etc.*
2. They create is peaceful, aesthetically pleasing environment.

3. Trees increase quality of life by bringing natural elements and wildlife habitats into urban setting.
4. Trees improve air quality, ameliorate climate, conserve water, preserve soil, maintain soil fertility and support wildlife.
5. Leaves absorb and filter the sun radiant energy, keeping environment cool in summer.
6. Trees lower the air temperature and reduce the heat in intensity of greenhouse effect by maintaining low level of CO_2.
7. They can create the impression of a well established place in new residential areas and reduce the raw unfinished look.

Classification of Trees

(A) According to Beauty of Plant Parts

(1) **For flowers-** *Bauhinia purpurea, Butea monosperma, Cassia marginata, Cassia nodosa, Logerstroemi speciosa, etc.*

(2) **For foliage-** *Alstonia scholaris, Araucaria cookie, Cupresssts funebris, Pinus longifolia, Salix babylanica, etc.*

(3) **For variegated foliage-** *Rosea morginata, Ficus benjamina, Ficus religiosa, Ornamental orange, Ficus elastic, etc.*

(4) **For fruits-** *Mangifera indica, Averrhoa carambola, Emblica officinalis, Tamarindus indica, etc.*

(5) **For fragrance-** *Magnolia grandiflora, Micholia alba, Plumeria rubra, Plumeria alba, etc.*

(B) According to Climate Condition

(1) **For moist areas-** *Alastonia scholaris, Cassia javanica, Lagerstroemia speciosa, Putranjiva roxburghaii, etc.*

(2) **For marshy area-** *Lagerstoemia speciosa, Salix babylonica, etc.*

(3) **For dry area-** *Delonix regia, Butea monosperma, Tectona grandis, Cassia fistula, Terminalia arjuna, etc.*

(4) **For arid area-** *Cassia siamia, etc.*

Purpose of Growing

A. **Specimen trees-** Such trees are planted singly for attractive shape, beautiful foliage, flowers or for drooping branches which reflect humbleness, *e.g., Cassia fistula, Plumeria alba, Magnolia grandiflora, Salix babylonica, Ficus elastica, etc.*

B. **Shade trees-** Such trees have mostly round canopy or umbrella crown. Leaves are so dense that no or very little sunlight is allowed underneath them, *e.g., Azadirachta indica, Alstonica, Ficus religiosa, Ficus bengamina, Mangifera indica, etc.*

C. **Ornamental trees-** Ornamental trees are those trees which are planted for enhancing the aesthetic value, fruit, stem, bark, habit and overall frame work are responsible for increasing the aesthetic value, *e.g., Putranjiva roxburghii, Salix babylonica, Saraca indica, Terminalia arjuna, Ravenala medagascariensis, Plumeria alba, Jacaranda mimosifolia, Grevillea robusta, etc.*

D. **Flowering trees** -These trees produce colourful flower and are planted for there beautiful flower, *e.g., Bauhinia variegata, Casssia fistula, Delonix regia, Butea monosperma, Plumeria alba, etc.*

E. **Trees for avenue and road area-** Avenue trees are planted along road side. Generally the avenue trees are planted for shade and flowers, *e.g., Cassia fistula, Jacaranda accutifolia, Ficus infectoria, etc.* Avenue trees are planted in different way-

   1. **One kind of flowering on both side**: *Bauhinia variegata, Gravellia robusta, Cassia fistula, Lagerstroemia speciosa, etc.*

   2. **Two kind of flowering tree blooming at one time on both side**:

      Gravellia robusta (yellow)- *Jacaranda accutifolia* (blue)- *Gravellia robusta* (yellow)

      Cassia fistula (yellow)- *Delonix regia* (red) - *Cassia fistula* (yellow),

      Cassia fistula (yellow)- *Cassia nodosa* (pink)- *Cassia fistula* (yellow)

   3. **Two kind of flowering tree blooming at different time on both side**:

      Bauhinia triandra (purple in November)- *Spathodea campanulata* (red in April)-*Bauhinia triandra* (purple in November)

      Cassia fistula (yellow in May)- *Jacaranda accutifolia* (blue in April)- *Cassia fistula* (yellow in May)

      Gravellia robusta (yellow in April)- *Bauhinia variegata* (purple in March) *Gravellia robusta* (yellow in April)

F. **Shady trees on both side**- Tree that are planted under avenue are *Cassia fistula, Cassia siamea, Delonix regia, Butea monosperma, Putranjiva roxburghii, Jacaranda mimosifolia, Terminalia arjuna, etc.*

G. **Screening purpose-** When tall upright tree are planted very close to give and ultimate look of curtain and screen, *e.g., Eucaliptus* species, *Populor species, Polyanathia longifolia, etc.*

H. **For fragrance flower-** *Terospermum acerifolium, Plumeria spacies, Magnoia grandiflora, etc.*

I. **For checking air pollution-** Industries are major source of pollution and hence in such areas deciduous tree or trees having thick shining leaves will be more successful, *e.g,. Morus species, Ficus infectoria, Ficus religiosa, Populer hybrid, Anthocephalus cadamba, etc.*

Description of some Important Flowering Trees

Sl. No.	*Common Name*	*Botanical Name*	*Family*	*Flower Colour*	*Flowering Time*	*Method of Propagation*	*Remark*
1	Amaltas the Java cassia	*Cassia fistula* *Cassia javanica* *Cassia nodosa* *Cassia siamea*	Fabaceae	Golden Yellow Pink PinkYellow	April-May May- June May- June May- June	Seeds	Medium sized to tall Beautiful flower Medium sized tree Medium sized beautiful Flower avenue tree
2	Gulmohar	*Poinciana regia syn. Delonix regia*	Fabaceae	Orange/Scarlet	April-May	Seed	Flower cover the whole tree
3	Pride of India	*Lagerstroemia flosreginae*	Lythraceae	Purple mauve	April-May	Seeds	Medium size tree
4	Blue gulmohar	*Jacaranda mimosifolia*	Bigoniacea	Blue mauve	March-May	Seed/semi-hard wood cutting	Elegant, deciduous tree
5	Bottle brush	*Callistemon lanceolatus*	Myrtaceae	Scarlet	March-Aug, Sep	Seeds layers	Small drooping branches flower born in spikes
6	Flame of forest (Palash)	*Butea monosperma, B. monosperma* var. *tutee*	Fabaceae	Scarlet orange yellow	April-May	Seed	Medium size, rough trunk deciduous
7	Kachnar	*Bauhinia purpurea, B. alba*	Fabaceae	Purple deep pink white	November	Seed	Medium size
8	Champa	*Michelia champaka*	Magnoliaceae	White/creamy yellow	April-May/Sep-Oct	Seed/grafting	Medium size long leaves
9	Pangri	*Erythrina indica*	Fabaceae	Scarlet red	Feb-April	Cutting	Fast growing, densely branched
10	Madar tree	*Gliricidia maculata*	Fabaceae	Pale Pink flower	Feb-March	Seed	Good looking, medium used for shading cocoa
11	Amherstia	*Amherstia nobilis*	Fabaceae	Vermilion yellow tip	Feb-May	Seed/Layering	Medium, beautiful tree
12	Indian lilac	*Lagerstroemia indica*	Lythraceae	Scarlet orange red	May-Aug	Seed	Small tree
13	Himanchampa	*Magnolia grandiflora*	Magnoliaceae	Pink white	April-May	Layering	Evergreen tree

Sl. No.	Common Name	Botanical Name	Family	Flower Colour	Flowering Time	Method of Propagation	Remark
14	Harsingar/shiuli	*Nyctanthes arbor-tristis*	Oleaceae	White with orange red tube	Sept- Nov	Seed/cutting	Sweet scented flower open at night and fall at day time
15	Akashneem	*Millingtonia hortensis*	Bignoniaceae	Silvery white		Seed/root suckers	Quick growing straight 20 m.
16	Yellow gulmohar	*Peltophorum inerme*	Fabaceae	Yellow	Feb-May/Sep-Aug	Seed/cutting	Beautiful ornamented tree
17	Temple tree/ pagoda tree	*Plumeria alba*	Apocynaceae	White	March-April/ July-Aug	Cutting	Evergreen tree
18	Chameli, Gul-e-chin	*Plumeria rubra*	Apocynaceae	White-rose		Cutting	Low growing 6-8 m tall tree
19	Sita ashok	*Saraca indica*	Fabaceae	Orange	April-June	Seed	8-10 m tall evergreen tree
20	Fountain tree	*Spathodea campanulate*	Bignoniaceae	Scarlet orange/ crimson	Feb- March	Seed/root-suckers	20-25 m tall ornamental
21	Tecoma	*Tecoma argentea*	Bignoniaceae	Bright yellow	Feb- March	Seed	5-8 m tall tree

Description of some Important Foliage Trees

Sl. No.	Common Name	Botanical Name	Family	Flower Colour	Flowering Time	Method of Propagation	Remark
1	Neem	*Azadirachta indica*	Meliaceae	Small white	April	Seed	12-18 m tall tree
2	Christmas tree	*Arucaria cookie*	Pinaceae	-	-	Seed	Ornamental tree
3	Siris	*Albizzia lebbek*	Fabaceae	-	-	Seed	Fast growing spreading tree
4	Jhau	*Casuarinas equisetifolia*	Casuarinaceae			Seed	10-20 m tall tree
5	Shisam	*Dalbergia sissoo*	Fabaceae			Seeds cutting	15-20 m tall tree
6	Banyan	*Ficus benghalensis*	Moraceae	Crimson berries	Aug-Sep	Seeds/cutting	15-20 m tall tree
7	Pakur	*Ficus infectoria*	Moraceae			Seeds/cutting	15-20 m tall evergreen spreading tree

Sl. No.	Common Name	Botanical Name	Family	Flower Colour	Flowering Time	Method of Propagation	Remark
8	Pipal	*Ficus religiosa*	Moraceae	Creamy white		Seed	20-25 m tall huge tree
9	Indian rubber plant	*Ficus elastica*	Moraceae	White, pink cream		Air layering	10-12 m quick growing tree
10	Silver oak	*Grevillea robusta*	Proteacae	Golden yellow	April-May	Seed	15-20 m evergreen tree
11	Mahua	*Madhuca longifolia*	Sapotaceae	Creamish white	March-April	Seed	15-20 m tall deciduous tree
12	Maulsari	*Mimusops elengi*	Sapotaceae	White	April-July/Sept-Nov	Seed	10-15 m tall dense evergreen tree
13	Polyalthia	*Polyalthia longifolia* *P. pendula*	Annonaceae	Golden yellow	July-Oct	Seed	10-15 m tall dense evergreen tree
14	Chir	*Pinus longiflolia*	Pinaceae			Seed	15-20 m tall needle like lives
15	Putranjiva	*Putranjiva roxburghii*	Euphorbiaceae	Yellow	March-May	Seed	10-12 m tall dense evergreen tree
16	Poplar	*Populous deltrolides*	Saliceaceae			Cutting	Fast growing, deciduous 5-10 m tall
17	Karanj	*Pongamia glabra*	Leguminosae	Lilac	April-May	Seed/cutting	5-10 m deciduous tree
18	Arjun	*Terminalia arjuna*	Combretaceae	Yellowish white	March-June	Seed	15-20 m tall evergreen avenue tree
19	Morpankhi	*Thuja occidentalis*	Pinaceae	White		Seeds	5-8 m tall, foliage is fern like
20	Travelers tree	*Ravrenala madagascariensis*	Musaceae			Seed	3-4 m tall, banana like leaf
21	Juniperus	*Juniperus chinenss*	Pinaceae			Seed	5-10 m tall, hardy, dense, pyramidal decorative tree
22	Chalta	*Dillenia indica*	Dilleniaceae	Large white	July	Seed/stem cutting	8-20 m tall, slow growing evergreen tree
23	Chitwan	*Alstonia scholaris*	Apocynaceae	White- greenish	April-May	Seed	Strong small during night, 6-10 m tall tree

Planting Care and Management of the Trees

1. **Climatic factor-** Selection of trees will depends upon tree veiling of locality and trees of other climatic condition may or may not do well or latter decline. This depends upon adaptability of individual types. Tree of tropical climate like *Delonix regia*, *Nauclea cadamba*, decline late on in subtropical climate and do not grow well in tropical region.
2. **Soil factor-** In general trees are available to varying soil condition but specific trees excel in particular condition. In alkaline and saline soil trees that can be grown successfully is, *e.g., Cassia fistula, Casurina equisetifolia, etc.*
3. **Care and management-**
   a. **Planting of trees-** Rainy season (June-July) is the best time of planting tree however in North India planting can be done during January-February. 1-2 year-old plants are most suitable for planting because the less chance of plant mortality.
   b. **Preparation of ground-** Pit should be done 60×60×60 cm in site and 10-15kg FYM and 20-25 gm of 5 per cent insecticide is used in soil.
   c. **Maintenance and care of trees**
      - ☆ Staking
      - ☆ Fencing and tree ground
      - ☆ Irrigation
      - ☆ Gape filling
      - ☆ Training and pruning.

Chapter 11

Climbers

Climbers is the group of plants which have weak stems and ability to climb up the support with the help of modified organs for sunlight and air. Twiners differ from climbers in the way that they do not posses such modified organs but twine around the support, cover it and reach the top. Climbers are woody or herbaceous plant which climb up the trees and other tall object with the help of support of their special modified organ, such as tendrils, root, thorns, rootlets, hooks, *etc.*

Tendrils- *Bigonia gracilis, Glimatis panniculata, etc.*

Thorns- *Baugainvillea species,* Climbing *roses, etc.*

Root lets- *Ficus repens, etc.*

Hooks- *Gallium aparine, Rubus australis, etc.*

Climbers and twiners are important group of the plants. They add the beauty colour or fragrance in the garden and artificial structure like garden wall, topiary, arches, pergola, *etc.* can be well developed with the help of climbers.

Different Modified Organs in Climbers

Tendrils- A specialized stem leaf or petioles with a thread-like shape that is used by climbering plants.

Leaf tendrils- In weak stemmed plant leaf or a part of leaf gets modified into green thread-like structure is called leaf tendrils.

Stem tendrils- Stem when gets modified into green thread-like leaf less structure is called stem tendril.

Thorns- It is a stiff shaped pointed woody projection on the stem which helps plant to climb up.

Climbing roots- These are aerial adventitious roots that help weak stem to climb on a support.

Clinging roots- These roots fix epiphytes on the bark on the supporting tree or pillar.

Rootlets- A small or fine branch of root by which weak stem plant climbs up the support.

Twiners- These are plants which are climbed by coiling around the support, a phenomenon known as circumnutating. They use their young shoot to twine around there climbing aid as they do not develop their climbing organs.

Ramblers and stragglers- They are plants which fail in their attempt to climb but somehow manage to support themselves over the trunk, stems or branches of other plants.

Latex- A milky fluid found in many plants which help them to climb up the support, *e.g., Ficus pumila.*

Importance of Climbers

- ☆ Beautify the surrounding.
- ☆ Covering slope, ground, ugly object and site.
- ☆ Create privacy.
- ☆ Provide feature in garden.
- ☆ Gives attractive view on trend over tree.
- ☆ Making topiary.
- ☆ Provide background for annuals and herbaceous border.
- ☆ Provide shade when trend over pergola.
- ☆ Provide fragrance to the surrounding.
- ☆ Supplementary.

Properties of Climbers

- ☆ Growth of shoots is often extremely rapid.
- ☆ Long internodes are produced for very rapid elongation. These often have sensitive contact with any support or solid object.
- ☆ Commonly there is a long delay in development of large leaves until the stem or cylindrical axis becomes wrapped around a support.
- ☆ Woody stems are very feasible to permit bending, twisting and coiling. They are fairly longer if pulling on two ends but have very little stems when compressed.
- ☆ If the supporting tree fails then the entire plants come-down however these climbers often have a great ability to survive and re-sprout.

Classification of Climbers

(A) According to Beauty of Plant Part

1. **Flowering climbers-** *Allamanda catharitica, Lonicera japonica, Antigonon lactopus.*
2. **For foliage-** *Asparagus plumosus, Ficus pumila, Piper ornatum.*
3. **Flowering and foliage-** *Artabotrys uncinatus, Ipomea quamoclit.*

(B) According to Nature of Climber

1. **Annuals-** Sweet pea, Purple bells, Morning glory, Passion flower.
2. **Perennials-** English evy, Virginia creeper.

(C) According to Situation

1. **Situation for partial shade condition-** *Clerodendrum splendens, Lonicera japonica.*
2. **Situation for sunny condition-** *Antigonon leptopus, Campasis grandiflora, Quisqualis indica.*
3. **Situation for shade and indoor condition-** *Asparagus plumosus, Monstera deliciosa, Philodendron* species.
4. **Situation for screening wall-** *Ficus plumila, Bignonia unguis cati.*

(D) According to Rate of Growth

1. **Heavy climbers –** The climbers that have fast and good growth and also produce profuse flower. These climbers are preferred to grow in huge area, *e.g., Bauhinia vahlii, Beaumontia grandiflora, Thunbergia grandiflora, etc.*
2. **Light climbers-** These climbers have less growth and spreading habit. These are well suited for less area, *e.g., Bignonia unguiscati, Ipomoea species, Tecoma jasminoides, etc.*

(E) According to Fragrance Flowers

1. **Fragrance climbers-** *Jasminum grandiflorum, Lonicera japonica, Solandra grandiflora, etc.*

Criteria for Selection of Climbers

- ✰ Amount of sunlight require
- ✰ Habit
- ✰ Spread
- ✰ Texture (fine, medium and coarse)
- ✰ Leaf retention duration

- ☆ Foliage colour (emergence, mature and fall)
- ☆ Flower colour (shape, size and fragrance)
- ☆ Flowering season and peak flowering duration
- ☆ Fruit (colour, shape, size, season and duration)

Purpose of Growing Climber

- ☆ **For screening purpose-** These climbers are evergreen, easy to grow and provide privacy in garden, *e.g., Bignonia unguiscati, Ficus pumila.*
- ☆ **For arches and pergola-** Climbers covering arches and pergola will provide vertical interest outside, *e.g., Salondra grandiflora, Pyrostegia venusta.*
- ☆ **For wall or trellis-** Climbers covering walls or trellis not only screen unsightly areas of the garden, but also brighten up bare walls, *e.g., Passiflora laurifolia, Ficus pumila, Bignonia unguiscati.*
- ☆ **For pot culture-** Light climbers are those having bushy growth are suitable for planting under pot culture, *e.g., Bignonia purpurea, Climatis flammaula, Hoya carnosa.*
- ☆ **For porches-** Climber are for creating privacy and beautify the porches, *e.g., Clematis panniculata, Clerodendron splendents, Wisteria sinensis.*
- ☆ **For covering slopes-** Such climbers are used for covering bare patches of soil around tree or shrub, *e.g., Thunbergia grandiflora, Lonicera japonica.*
- ☆ **For making topiary-** Climbers with repeated pruning and flexible vegetative growth are suitable for making the topiary, *e.g., Bignonia species, Clerodendrum inerme.*
- ☆ **For hanging baskets-** Plants are suitable for planting in container, *e.g., Bougainvillea, Clematis, Hedera hellix.*
- ☆ **For ornamental fruits-** These are the climbers that are properly grown for their beautiful fruits, *e.g., Dioscoria deltoidea, Hedera nepalensis.*

Planting of Climbers

Climbers can thrive well on any soil. The basic requirement is the soil should be fertile, deep, well drained, with good water holding capacity. Planting should be done in a pit of 60×60×60cm size. Before planting it must be refilled with 10-15 kg of well rotten FYM and 10 kg Falidol powder for planting evergreen climber, rainy season is preferred, *i.e.,* July-Sep. Although it can also be planted during February-March. The deciduous climbers should be planted during winter, *i.e.,* February-March.

Care and Management

The regular watering is required after planting the climber. Weeding and hoeing must be done as and when required. During subsequent year pruning is essential to keep the climber in limit and in desired shape.

Description of some Important Climbers

Sl. No.	Common Name	Botanical Name	Family	Flower Colour	Flowering Time	Method of Propagation	Remark
1	Rangoon creeper (madhu- malati)	*Quisqualis indica*	Combretaceae	White and pinkish	Most part of the year	Cutting, layering	Flowers are produced in drooping branching
2	Jhumkolata (passion flower)	*Passiflora edulis, P. alba*	Passifloraceae	White, pink, purple	June-Nov	Suckers, layering	Vigorous, hardy
3	Madhabilata	*Hiptage beneghalensis*	Malpighiaceae	White with yellow	Dec-Feb	Seeds/layering	Sweet scented evergreen
4	Allamanda	*Allamanda cathartica* *A. Hendersonii*	Apocynaceae	Yellow	April-July	Cutting, layering	Easy to grow
5	Coral creeper	*Antigonon leptopus*	Apocynaceae	Orange yellow	Most part of the year	Seed, cutting, layering	Tuberous rooted, quick growing
6	Duck flower	*Aristolochia elegans* *A. grandiflora*	Aristolochiaceae	Rose	April-June	Seed, cutting, layering	Flower emit repelling odour
7	Dutchman's pipe	*Aristolochia adenocalymma*	Bignoniaceae/ Aristolochiaceae	White purple brown	March-June	Layering, cutting	Emit a garlic like smell
8	Trumpet creeper	*Campsis radicans*	Bignoniaceae	Yellow pink mauve	July-Aug	Suckers, cutting	Trampet-shaped flower with aerial rootlet
9	Clerodendron	*Clerodendron splendens*	Bignoniaceae	Red deep orangre	Dec-Feb	Suckers, layering	Beautiful climber
10	Golden shower	*Bignonia venusta*	Bignoniaceae	Yellow	Feb-June	Layering, cutting	Tabular flower grown on compound wall
11	Purple wreath	*Petrea volobilis*	Verbenaceae	Purple	Feb-April	Suckers, cutting Layering	Star-shaped flower
12	Juhi	*Jasminum auriculatum*	Oleaceae	White	April-Sept	Cutting	Scented flower
13	Safe bel bridal bouquet	*Porana paniculata*	Convovulaceae	White	Aug-Oct	Cutting, layering, seed	Head-shaped leaves used for screen
14	Heavenly	*Thunbergia grandiflora*	Acanthaceae	Blue with yellow	Feb-Aug	Seed, cutting	Used for covering wall
15	Vernonia	*Vernonia elaeagnifolia*	Compositae	White	July-Aug	Seed, cutting	Used for screening
16	Money plant	*Pothas aurens*	Areceae	Orange, yellow		Cutting	

Chapter 12

Ferns and Palms

Ferns

Ferns are flowerless plants that has feathery, leafy fronds. These are reproduced by spores released from the under sides of the fronds. A wide range of beautiful plants exists among ferns which are very popular worldwide. Ferns are grown for their beautiful foliage and their gracefulness. Most of the house ferns are native of tropical regions and are ideal for indoor planting, as they can grow well under shaded or semi-shaded situation. Many species of ferns are native to India. Some of them can be easily grown while others are much difficult, but can be grown well under preferable suitable growing media and proper management.

Some Important Species of Ferns

1. **Maiden hair fern (*Adiantum capillus-veneris*)** - This is the shade loving easily growing fern require plenty of water for its growth. It is propagated by division or from spore.
2. **Asplenium (*Asplenium bulbiferum*)** - It is a shade loving fern that require much water for its proper growth. Plantlets are used for its propagation. It is a good house plant
3. **Holly fern (*Cyrtomium falcatum*)** - It is suitable for both shade and partial shade. This is the indoor plant, commonly propagated by division or from spore.
4. **Microlepia fern (*Microlepia strigosa*)** – It is a robust fern with delicate finely cut triangular fronds. It is useful for landscaping of shaded area of the garden.

5. **Lemon button fern (*Nephrolepsis furcans*)** – It is a terrestrial, perenial, medium sized herbaceous fern. Leaves are compound, unipinnate glabrous and evergreen.
6. **Golden rod fern (*Phymatodes scolopendria*)** – It is an epiphytic, medium sized, perenial, herbaceous fern. Fronds are deeply pinnatified, glabrous and fertile with simple biforked shinning leaves.
7. **Staghorn fern (*Platycerium bifurcantum*)** - It is easy to grow and require partial shady condition. It is propagated by offsets.
8. **Rabbit foot's fern (*Nephrolepis exaltata*)**- It is a terrestrial, perenial, medium sized, deciduous and herbaceous fern. It is propagated by stolon and rhizome.

Care and Management

Soil

Fern grows well in light acidic soil. This is a potted plant and require generally medium sized pot with a good soil mixture having 1 part sand, 2 part leaf mould, 1 part soil, 1 part small piece of charkol, 1 part crushed lime and few small piece of broken bricks.

Irrigation

Watering is very essential for growing beautiful fern. Generally fern is not requiring sprinkling of water, for better growth moisture should maintain in soil medium. Excess watering turn browning in their foliage.

Fertilizers

Nutrition is the very important for fern growth and development. Liquid nitrogen response better and they can apply at frequent interval. Blood meal is very good for fern as it promote development of nice colour in leaves. It should be applied about teaspoonful once in every 6-8 week.

Container

Earthen pot are cheaper and being porous provide better aeration to the soil. Plastic pots and bowl are not suitable for growing moisture loving fern. The shape of the pots may be round, oval, cone, rectangular and squire or any shape. Pot should have drainage hole at the bottom.

Palms

Palms are group of the plants which have great ornamental value and are planted in the garden for special purpose as most of species have unbranched single shape. Tufted leaves at the end of stem create most picturesque effect. An unbranched, evergreen tree of tropical and warm regions with a crown of very long feathered or fan shape leaves and typically having old leaf scars forming a regular pattern on trunk. Plants are native of tropics and belong to Palmae or Arecaceae family. They are varied in forms, size and habit of growth. Palms do not increase in thickness and also do not show any distinction in bark wood and pith. Flowers are produced on spikes that hanged gracefully in many species. Flowers may be bisexual or unisexual borne either on the same plant or on separate plant. Whereas, some palms are polygamous.

Feature of Palms

1. Palm is easy to cultivate.
2. The palm is of great ornamental value.
3. An ideal planting material in landscaping.
4. They can be planted under trees as interesting specimens or groups.
5. Most palms are commonly used in avenue planting.
6. They are of great economic values (provides shelters, food, clothing, fibers, timbers, oil, sugar, starch, *etc.*).
7. They can be planted as focal point in the landscape or in the center of the lawn.

Based on the Trunk Characteristic Palms are Divided into Five Groups

1. **Solitary**- These single trunk stemed growth habit is very common and is characteristic of many palms. The are cultivated for ornamental and economic purpose and great variability exists in both height and diameter of solitary palms, *e.g., Jubaea chilenis, Ceroxylon alpinum.*
2. **Clumping-** Characterized with multiple trunk and are also quite common, from a common root system palms produces suckers (Basal off shoots) at or below ground level. These suckers grow to maturity and replace the oldest stem as they die, *e.g., Phoenix dactylifera.*
3. **Aerial branching**- These palms are unusual and only found naturally. Branching occur by equal forking at the growth point and may occur as many as five times. Because of sub-lethal damage to the growing point by insect or a physical force. Lighting aerial branching, *e.g., Cocos nucifera, Laccosperma* species.
4. **Subterranean branching**- It occurs by at least two process—either farming dichotomous branching or lateral. Lateral plant produces subterranean branching. It is produced by separating and transplanting individual branching, *e.g., Nypa fruticans.*
5. **Trunk less-** They have reduced trunk and referred as acquiescent. These are climbing type, *e.g.,* Calamus.

Criteria for Selection of Palms

- ☆ Habit
- ☆ Habitat
- ☆ Form

Description of some Important Palms

Sl. No.	Common Name	Botanical Name	Origin	Plant Height (cm)	Remark
1	Areca palm/Butter Fly palm/Cane palm	*Areca lutescens*	Madagascar	800	Stem yellowish, cylindrical, graceful, midrib, glossy yellow green
2	Panama Hat palm	*Carludovica palmata*	Central America to Bolivia	250	Leaves are fan shaped
3	Fish Tail palm	*Caryota mitis*	Andaman, Iceland	700	Leaves are wedge shaped or fish tail like leaflets
4	Chameadorea	*Chameadorea elegans*	Mexico	250	Beautiful spineless small palm
5	Livistonia/Chinese Fan palm/Chinese Fountain palm	*Livistona chinensis*	China and Japan	900	Leaves bright green, fan like. Suitable for avenue and pots
6	Phoenix/Date palm	*Phoenix roebelenii*	Tropical Asia and Africa	900	Very graceful, handsome dwarf palm and pot plant
7	Rhaphis/Bamboo palm	*Rhaphis excelsa*	South China and Japan	150-300	Leaves leathery, glossy green. Suitable for shade garden and as pot plant
8	Thrinax/Fan palm	*Thrinax parviflora*	Jamaica, Cuba and Florida	800	Leaves almost circular in shape, green colour on both side with yellow rib and pot plants
9	Cardboard palm Zamia Palm	*Zamia furfuracea*	Mexico	150	Living fossil plants, thick leathery leaf give a cardboard like feeling on being rubbed
10	Bismarck palm	*Bismarkia nobilis*	North America	200	Leaves greenish to silvery green
11	Royal palm/Bottle palm	*Oreodoxa regia/Roystonia regia*	North America	200	Rigid swollen stem with bulge in middle (bottle shape), terminal leaves
12	Queen palm	*Arecastrum romanzoffianum*	South America	800-1000	Arching, feather-like glossy leaves, green

- ☆ Height
- ☆ Spread
- ☆ Shape
- ☆ Structure
- ☆ Leaves
- ☆ Flowers

Propagation and Management

Palms are propagated by seed and sucker. Seed propagation is very slow and seeds vary in size from the size of pea to the size of coconut. Seeds are covered with thick coat which make germinate very slowly. Palm seed have to be harvested at they are ripen and or to be sown immediately for obtaining best quality seedling and high percentage of germination.

The seedling should be planted in small pots which are sufficient to accumulate the root. At the time of reporting the flashy large root should not injured. Deep planting should be avoided. Regular supply of water should be given in regular interval. Good drainage conduces to healthy plant. At the time of potting mixture good garden soil, coconut peat, well rotten compost, coarse sand, *etc.* in balance ratio.

Chapter 13

Bulbous Plant

These ornamental plants which have specialized modified underground stem structure to overcome the unfavourable environmental conditions. They are known as a bulbous plant. Botanically plants propagating themselves through modified underground stem are grouped into 4 groups–

1. **Bulbs-** Tuberose, Lillium, Tulip and Amaryllis, *etc.*
2. **Corms-** Gladiolus, Alocacia.
3. **Rhizomes-** Canna, Iris, Bamboo.
4. **Tubers-** Dahlia, *Begonia* species.

1. Bulbs

Bulbs is a specialized underground organ consisting of a short, fleshy, usually vertical axis (Basal plate) bearing at its apex a growing point or flower primordial enclosed by thick fleshy scales.

- ☆ Tunicate and concentric, *e.g.,* Tulip
- ☆ Non-tunicate and non-concentric, *e.g.,* Lillium.

2. Corms

It is a swollen base of stem axis growing vertically with distinct, node, internodes, terminal and lateral buds enclosed with thin papery scale-like leaves.

3. Rhizomes

It is a specialized underground stems in which the main stem axis grows horizontally at or just below the ground surface.

4. Tubers

It is specialized kind of swollen underground modified stem structure with a terminal and several lateral eyes which function as storage organ.

a. Tuberous Root

Tuberous root include several type of stem structure with thickened tuberous growth that function at storage organs. The buds farmed at the crown proximal end and the root at digital end, *e.g.*, Dahlia, Sweet pea, Tapioca, *etc.*

b. Tuberous Stem

It is the stem structure produced by enlargement of the hypocotyls of seedling plants and eyes are farmed on all through the swollen organ, *e.g.*, Begonia.

Care and Management

Climate

In general cool weather, high atmospheric humidity and moist soil condition or congenial for the manufacture of food, better production of high quality flower and bulbous. According to the climate bulbous require two season–

i. **Warm season-** Canna, Dahlia, Tuberose.

ii. **Cool season-** Gladiolus, Daffodils.

Soil

Soil should be sandy loam or loamy sand in texture which is best for better production. It should be moderately fertile with good water holding capacity and with pH range from 7-8.

Propagation

Bulbous plant can be propagated by seed, terminal cutting, division and bulblets. For bulblet it takes about 1-2 years for growing into a full bulb. In case of division tuber, rhizome and corm are divided carefully and planted in the soil. Seeds and terminal cutting is the most common method of propagation. Dahlia and Rex begonia are propagated by terminal cutting and seed propagation is generally used by breeder for creating new variety.

Manuring

Well rotten farm yard manure @ 5-6kg/m^2 should be incorporated at the time of soil preparation. Application of 20g each of phosphorus and potash per m^2 is also found beneficial at the time of bulb planting. Nitrogen @ 40g/m^2 in two split dosage should be applied.

Planting

In north India planting of bulbous plant like gladiolus, dahlia, iris, *etc.* can be done from mid Sep-mid Nov. Tuberose and Zephyranthes are separated and planted in Feb-March and canna is planted in July with the onset of rain. Depth of planting of bulb is 7-10 cm and distance between row and plants are 30×20 cm.

Irrigation

For better growth and development of the bulbous plant soil moisture should be maintained by regular light irrigation. Generally crops should be irrigated 7-10 days interval in winter season and 5-7 days interval in summer season.

Staking

Bulbous plant like gladiolus, dahlia, lilly, *etc.* produce flower on long stem and thus need support otherwise spike may bend down and break with strong wind. Bamboo stem can be used for providing support.

Harvesting

The stage of harvesting are depend upon the marketing for local market. Gladiolus and tuberose should be cut when lower 2-3 floret opens. Whereas, for distance market flower should be cut when lower 1-2 floret opens. Before harvesting the flowers plant should be irrigated.

Storage

Tender bulb like Gladiolus, Dahlia, Asphodel, *etc.* are to be dug from the soil 10-12 week after flowering has been finished. Before lifting the bulb, irrigation is stopped. After digging bulb are dried in shade for few days. Hardy bulbs are separated after 2-3 year and are again planted in the planting season. Bulb should be treated with 0.2 per cent Bavistine solution for 30 minute and there after storage.

Chapter 14

Cactus and Succulent

Cactus

It is a succulent plants with a thick fleshy stem which typically bears, spines, lacks leaves and has brilliantly colour flower.

Succulent

Plant that have part that's more than normal thickened and fleshy usually to retain water in arid climate and soil condition.

Important Features of Cacti and Succulents

1. Can tolerate unfavourable drought condition.
2. Have a waxy outer coated with wool.
3. Spines are present.
4. Contain efficient food and water storing structure.
5. In cacti areoles are present.
6. They thrive well in sunny situation and light lovers.
7. They have different attractive shape like animal, bird, rock, plants, *etc.*
8. Flowers are of brilliant colour and beauty.
9. Flowers vary in size from small to large.
10. All cacti belong to the family cactaceae.

Difference between Cacti and Succulents

Cacti	*Succulents*
Areoles present	Areoles absent
All cactus are succulents	All succulents are not cactus
Family – Cactaceae	Family – Cactaceae
Its nature is perennial	Succulents are perennial and annuals both
Leafless (except rose cactus)	Leafless
Dicotyledons	Dicotyledons
Seed—one seeded berry	Berry

Uses of Cactus and Succulents

1. **Outdoor growing-** In open air cultivation the plants develop faster, the shape also grows to its best and flower develop beautifully.
2. **Indoor growing-** There are fortunately a number of species that make well for less light then they are accustomed to and grow well under indoor climatic condition, *e.g., Pincushion cactus, Notocactus apricus, etc.*
3. **Bowl culture-** Any kind of bowl or dish, glazed or unglazed, China clay wooden or plastic material for growing cactus and succulents provided that if the depth is enough to hold the plant upright, *e.g., Opuntia imbricate, Opuntia cylindrica* (cacuts), *Crassula argentea* (succulents), *etc.*
4. **Cold greenhouse -** Most of the plants grow well in the cold frame or greenhouse besides many other that are too big or unsuitable in other way to induce, *e.g., Opuntia cylindrica, Crassula argentea.*
5. **As fragrant flower-** Some cacti produce sweet fragrance flower along with their beauty of plant, *e.g., Astrophytum sylvestris, Hylocereus extensus, Nycturus serprentine, etc.*
6. **For desert garden-** *Mammillaria cephalo, Cephalocereus senilis, etc.*
7. **For hanging basket-** *Aporocactus flogrelliformis* (cactus), *Kalanchoe, Euphorbia* (succulents), *etc.*

Areoles

Areoles are small light to dark colour bumps on cacti out off which grow cluster of spine. Areoles are important diagnostic feature of cactus and identify them as a family. Dusting from other succulents plant areoles represent highly specialized branches on cactus. Apparently they involve as abortive branches buds while there spine involve there westagious leaf. The development of areoles seems to have been an important element in the adaptation of cacti needs in desert ecology.

In botany areoles are small light to dark coloured bumps on cacti out of which grow clusters of spines. Some of the opuntiodeae have spine as well as glochids on their areoles some have only glochids. Spines are growing from areoles and hot directly from the plant stem.

Planting Care

Climate

Cacti are the widely adoptable crop and it can be grown in wide range of climatic condition. The extreme high and very low both temperatures are injurious for growth of cacti. Cacti and succulent are grown in specially semi arid or desert plant but some plants live in tropical regions. The plants also have resting period. During this period cacti need little care and watering at longer interval. Active growth start again during spring season and continue till autumn.

Soil

Cacti thrive well in porous and calcareous soil. Cacti are good for potted plants and potting mixture is very essential for their growth. The ideal potting mixture have 2 part garden soil, 1 part sand, 1 part leaf mould, 1 part well rotten manure, 1 part lime stone and 1 part charcoal. The small amount of bone meal is also beneficial. Cacti require more drainage than succulent so add more sand.

Propagation

Cacti and succulent plants are propagated through seeds, cutting, offsets and grafting.

(a) By Seeds

1. It looks longer time by seed as the growth rate of seedling is very slow.
2. Best time of seed sowing is mid March to June.
3. Seeds in a warm area without sunlight and cover with a plastic.
4. Germination takes place in 7-10 days. After germination move flat so plant receives more light and removes cover.

(b) By Grafting

1. Usually weakly grown species are grafted on strongly grown stock.
2. Best time for grafting is spring season or autumn season.

(c) By Cutting

1. **Stem cutting-** Cut stem and leaf system during the summer month. Cactus produces roots on tip section while succulent produces roots on stem nodes.
2. **Leaf cutting-** After cutting a leaf let it dry for a few days and places it in a sandy loam soil.

(d) Off shoots

Small plantlets are separated and some time divided from the stock plant. Succulent that produce offset include Aloe, Agave and Crassula, *etc.*

Potting Plants

1. Place one or two pieces of broken part over the drainage hole.
2. Place the plant inside a well, *i.e.*, deep enough to cover the root ball.
3. Fill the pot with soil mix until the root crown is covered but no more.
4. Water plant thoroughly so the soil is saturated each time and then allow drying before watering again.

Repotting

Best to repot in early spring of autumn season in the spring cactus will start growing and in the autumn season a repotted plant will adjust to the light and temperature change.

Light Requirement

1. Photo period (Light duration) will affect the stem colour as well as the formation of flower bud.
2. Flower buds farm keep the plant in the same location. Buds could be dropped if the plant is moved to a new location.
3. Low light may occurs the stem to turn light green or yellow and grow long stems (Etiolation- absence of light) in the direction of light.

Irrigation

Cacti and succulent do not need liberal watering. Judicious application of water is required for successful growing. In winter during rest period less watering is required. In summer frequent watering should be done for proper growth. Every time soil is to be drenched completely so that water comes out of the drain hole and over watering should be avoided.

Description of Important Cacti

1. Astrophytum (*Asrtophytum* species)

- ✰ Commonly known as star cactus or urchin.
- ✰ Bishops cap and goat horn cactus.
- ✰ Globular shapes.
- ✰ Areoles closely set along ribs.
- ✰ Ideal for small pots.
- ✰ Native to north Mexico.

2. Capricorne (*Astrophytum capricorne*)

- ☆ Cylindrical and tall growth up to 30 cm.
- ☆ 8-9 ribs with wooly areoles.
- ☆ Spines are thirsted.
- ☆ Native to north Mexico.

3. Cephaloceus (*Cephalocerus* species)

- ☆ Tall, columnar and cylindrical stem.
- ☆ Cover with long grayish white wooly areoles.
- ☆ Ribs are closely attached (*Cephalocerus senilis, Cephalaocerus fleemiensis*).
- ☆ Native to Mexico.

4. Cereus (*Cereus* species, *Cereus hexagonus, Cereus argentinensis*)

- ☆ Commonly call night flowering cactus.
- ☆ Tree sized cactus.
- ☆ Stem usually blue green.
- ☆ Flower white or pink.
- ☆ Native to Mexico and USA.

5. Cleisto Cactus (Root Stock) (*Cleisto cactus strausii*)

- ☆ Narrowing stem near the growing point.
- ☆ Stems are usually no more than one inch thick.
- ☆ Stem surface is hardy visible.
- ☆ Commonly used as root stock.
- ☆ Native to Bolivia.

6. Ferocactus (*Ferocactus* species) *Ferocactus covilliei* and *Ferocactus latispinus*

- ☆ Spines beautiful colours.
- ☆ Plants grow slowly.
- ☆ It contains fierce spines.
- ☆ Flowers are red yellow produce.
- ☆ Native to Mexico and USA.

7. Lobivia (*Lobivia famatimensis*)

- ☆ Commonly known as the cob-cac.
- ☆ Stem is globular and cylindrical.

- ☆ Ribs divided into tubercles.
- ☆ Spines stouts and directed upward.
- ☆ Flowers large, lemon yellow and games in days.
- ☆ Native to Argentina.

8. Opuntia (*Opunita* species)

- ☆ Very common cactus mostly planted for fencing purpose.
- ☆ Stem are globular or cylindrical and oval in shape.
- ☆ Demarcated by presence of flat and joined item.
- ☆ Commonly used as root stock.
- ☆ Native to Mexico and USA.

9. Pereskia (*Pereskia aculeate*)

- ☆ Large shape like cactus.
- ☆ Commonly called as lemon vine.
- ☆ Leaf large, well farm, fleshy, permanent and showy in appearance.
- ☆ Much branched having spine.
- ☆ It is used in root stock.
- ☆ Native to Tropical America.

10. Rebutia (*Rebutia* species)

- ☆ Small size cactus with free clustering habit.
- ☆ Flowers red to orange, pink, yellow, white fennel-shaped produce at site.
- ☆ Propagated by division of offsets.
- ☆ Native to Bolivia.

11. Zygo Cactus (*Zygocactus* species)

- ☆ Epiphyllous plants.

Description of Important Succulents

1. Aloe – Aloe Species (*Aloe aristata* and *Aloe humilis*)

- ☆ Family- Liliaceae.
- ☆ Leaves are often banded or contrasting colour.
- ☆ Plant varies in height.
- ☆ Native to South Africa.

2. Agave – (*Agave americana* and *Agave angestifolia*)

- ☆ Family – Sparagaceae.
- ☆ Also known as sanctuary plant.
- ☆ Leaves are sward shape often toothed and carried in roseum.
- ☆ Native to Mexico, North America.

3. Adenium (*Adenium obessum*)

- ☆ It has flashy and twisted stem.
- ☆ Grow well in pots (2m height).
- ☆ Flower colours are pink.
- ☆ Family - Apocynaceae.
- ☆ Native to Tropical America.

4. Heart Leaf Ice Plant (*Aptenia cardifolia*)

- ☆ Family - Aizoaceace.
- ☆ Pro-straight growth habit.
- ☆ Leaves are heart-shaped, fleshy, covered with papillae.
- ☆ Flower small shape purple red colour.
- ☆ Blaming during summer and rainy season.
- ☆ Native to South Africa.

5. Crassula (*Crassula ovata*)

- ☆ Family –Crassulaceae.
- ☆ Native to South Africa.
- ☆ Herbaceous or semi-herbaceous plants.
- ☆ Leafs opposite produce terminal lateral cymes.

6. Euphorbia (*Euphorbia bojeri*)

- ☆ Family - Euphorbiaceae.
- ☆ Native to Tropical Africa.
- ☆ Aothorny and flashy plant.
- ☆ Characterized by presence of milky juice.
- ☆ Flowers are yellow, greenish, yellow, red colour (also called cyathia).
- ☆ Propagated by separation of rooted offsets.

7. Kalanchoe (*Kalanchoe tomentosa* and *Kalanchoe beharensis*)

- ☆ Family— Carassulaceae.
- ☆ Native to Tropical Africa and America.
- ☆ Usually with branching and shrubbery parts.
- ☆ Leaves are smooth or felted.
- ☆ It grows fast and produce flower in cluster on long stock.

8. Lithops (*Lithops optica*)

- ✰ Family – Mesembatyanthemaceae (Lithioceae).
- ✰ Native to South Africa.
- ✰ It is the pot plant.
- ✰ Plant is grown singly or in cluster.
- ✰ Flower is short, yellow or white in colour.
- ✰ Propagated by seed and cutting.

9. Pedilantus (*Euphorbia tithymaloides*)

- ✰ Family – Euphorbiaceae.
- ✰ Native to North America.
- ✰ Popular for foliage beauty.
- ✰ Hardy in nature.
- ✰ Used for making hedge, topiary and grown in pots.
- ✰ Tall, branched, spineless and cylindrical.
- ✰ Leaves are oval-shaped small and dark-green colour.

10. Sedum (*Sedum nudum* and *Sedum compactum*)

- ✰ Family–Crassulaceae.
- ✰ Native to Europe.
- ✰ Herbaceous and slightly woody branch.
- ✰ Creeping in habit.
- ✰ Leaves alternate, flat and cylindrical in shape.

Chapter 15

Bonsai

Bonsai are miniature trees grown in pots. The aim of bonsai culture is to develop a tiny tree that has all the element of large tree growing in a natural setting. The texture of trunk, its look of age, the moss and the under planting in the container all contribute to the illusion of a miniature tree as it is seen in nature.

Bonsai is an art of growing dwarf ornamentally shape tree or shrubs in sallow container or tray. It is an art that expresses in miniature form of a nature tree. The word bonsai is comprises of two word, *Bon* means- A-tray or sallow container and *Sai* means- to-grow plants/tree. Bonsai is a Japanese term.

Factor Affecting Bonsai

- ☆ Containers/Pots (Cemented, plastic and wood).
- ☆ Selection of suitable plant species.
- ☆ Adequate sunshine.
- ☆ Suitable growing media.
- ☆ Careful trimming, training, pruning, wiring and reporting.
- ☆ Proper growth of root and trunk.

Criteria for Selection Plants

- ☆ Plant with small flower and fruit must be selected.
- ☆ Plant bearing flower on leaf less branches.
- ☆ Plant characters and it climatic requirements.
- ☆ Plants grow rapidly with good health.
- ☆ Ability of plant growing artificial condition other than its natural ones.

Classification of Bonsai

1. **Large bonsai**– This type bonsai are found 60 cm height.
2. **Medium bonsai**– Most popular bonsai are ranged between 30-60 cm height and easy to handle.
3. **Small bonsai**– Plants reached upto height of 30 cm. which in the mainly shrub and tree are selected.
4. **Miniature bonsai**– This develops upto height 15 cm. It is also called as Mame or Finger tip bonsai.

Principles of Bonsai

To produce realistic elusions of a mature tree look for plants with the following characteristic–

a. Short internodes are distance between leaves.
b. Small leaf and needles.
c. Attractive bark and root.
d. Branching characteristic for good wing farm.

Style of Bonsai

1. **Formal upright style or chokkan-** It grows straight tapered to the apex with clear trunk having alternately arranged branch, trained to pyramidical or spherical shape, *e.g., Ficus diversifolia, Cardia sebstena, Ficus triangularis* and *Ficus nitida, etc.*
2. **In formal upright style or moyogi-** It grows upward but irregular in movement. The informality can be achieved by planting in a little slanted position, *e.g., Callistemon, Lanceolatus, Casuraina equisetifolia, Junipers species* and *Punica garnatum, etc.*
3. **Slanting trunk style or shakkan-** In this style the tip of the tree is always kept upward and arrangement of branches is somewhat different from upright one, emphasis must be given on growing slant on the right or left of the root, *e.g., Callistemon danceolatus, Ficus diversifolia, Cinnamomum comphora* and *Punica granatatum, etc.*
4. **Cascade style or kengai-** In this style the plant grows upright and then it abruptly turn to grow downward. The tip of the cascading trunk may extraned beyond the bottom of the pot, *e.g., Juniper* species and *Verginia cripper, etc.*
5. **Semi-cascading style or han kengai-** In this trunk grows horizontally where the trunk line may be straight or curved branches arise from upper side of the trunk in very unusual pattern, *e.g., Juniperus horizontalis* and Mountain pine, *etc.*

6. **Multiple trunk style or kabudachi-** This is characterized by two or more trunk that also includes twin trunk, triple trunk and multiple trunk style. These are much sought but are scarcely available. Generally oval pot is used for planting, *e.g., Callistemon lanceolatus, Cinnamomum comphora* and *Punica granatum, etc.*
7. **Clump style-** When tree or shrub produce a series of shoot around there base it is called clump style. Trunk can be trained formal and informal style, *e.g., Fortunella japonica, Malpighia coccigera* and *Cinnamomum comphora, etc.*
8. **Broom style or hokidachi-** Plant both shrubs and trees having almost vertical branching from almost ground level and with little trunk, *e.g., Malpighia coccigera* and *Thuja orientalis.*
9. **Windswept style or fukinagashi-** This is a replier of a tree or a group of trees expose to prevailing. The trunk right from base is slanted on one side right or left keeping apex straight branches are trained to grow almost horizontally or slanted side of plant, *e.g.,* Junipers, *Casuarina equisetifolia, etc.*
10. **Rock grown style or ishisuki-** This style is comprises with the combination of rocks plants, *e.g., Salix* and *Babylonica*.
11. **Forest or group or yose ue-** This style comprising the planting of several or many trees of one species, typically an old number, in a bonsai pot.

Different Style of Bonsai Tree

Suitable Plant for Bonsai

a. Trees

Mangifera indica, Malus pumila, Bomboosa nigra, Ficus infectoria, Pinus dancyflora and *Putranjiva roxburghii, etc.*

b. Shrubs

Bougainvellea, Adenium, jatropha and *Hamelia patens, etc.*

Containers for Bonsai

- ☆ It should harmonize with plant to add beauty.
- ☆ It must have proper drainage hole.
- ☆ Colour of pot should match with style of plant.
- ☆ Size and shape must correlate the plant.
- ☆ It should be aesthetic and functional.

Potting and Repotting

For making bonsai the plant material may be seedling cutting, layering or grafting. Seedling should be first planted in small pot with the increase in age and size. This planting may be at 20-30cm prior to their transfer to the ornamental bonsai pot. Pots have green, blue or white colour fit well for deciduous species. In bonsai repotting actually the being of presenting the plant in its perfect style. It consists of removal of plant either from ordinary pot or from bonsai pot and replacing it to either first time in bonsai pot or in same ornamental bonsai pot. The best season for repotting is earlier spring or monsoon. Evergreen plants require repotting after 3-5 year.

Wiring

Wiring used for training the branches and the trunk of bonsai to a desired shape and bending in any direction. Two type of wire are used–copper and aluminum. The wire thicknesses vary as per thickness of branch or trunk desire to be given shape. While wiring branches keep wire at 45° angles to branch for best result, wire should be removed after few months when it is starts titrating into bark of plant.

Training and Pruning

- ☆ Training and pruning is done for imparting dwarfing and desirable shape of bonsai.
- ☆ Pruning keeps the plant within a height limit.
- ☆ Increase branches in a desire direction.
- ☆ It trained the root system to fit in sallow container.
- ☆ It controls the root system of matured plant.

- ☆ Proper training provides the best shape to the plant.
- ☆ In one trimming not more than 1/3 portion of root and shoot are cut.

Care and Management

Bonsai care and management requires techniques and tools that are specialized to support the growth and long-term maintenance of trees in small containers. Though bonsai trees are more delicate, compared to the average indoor plant, a few basic rules should enable anyone to take care of its tree properly. Most importantly are soil mixture, watering and fertilization, *etc.* Soil is important to supply your trees with nutrients, but it also needs to drain properly, provide enough aeration and retain water. Balanced soil mixture is most important for bonsai plant.

The soil mixture are prepared with 3 parts of loam soil, 3 parts of cow dung manure, 3 parts of leaf mould and 2 parts of brick pieces or sand. The most important part of taking care of your bonsai trees is watering. How often a tree needs to be watered depends on several factors (like species of tree, size of tree, size of pot, time of year, soil-mixture and climate), indicating that it is impossible to say how often you should water bonsai. However, understanding a few basic guidelines will help you to observe when a tree needs to be watered. Fertilizing regularly during the growth season is crucial for your bonsai to survive. Normal trees are able to extend their root system looking for nutrients. However, bonsai are planted in rather small pots and need to be fertilized in order to replenish the soil's nutritional content.

Defoliation is done where all the leaves are removed from tree to force new flush. The best time to practice defoliation is July to August or February to March, but it largely depends on plant species. Similarly leaf pinching and shoot tip pinching is generally done in plants which are grown for flowering beauty. Therefore the application of insecticides and pesticides at regular intervals must be done to keep off pests and diseases.

Chapter 16

Plant Growth Regulators

The term plant growth regulators is relatively new in use. In earlier literature these were mention as Hormones. A hormone is a Greek word derived from *Hormao* means Stimulate. Thimonn (1948) suggested the use of term phytohormone in place of hormone of plant.

Growth

Growth is the quantitative increase in the plant body, *e.g.,* increases in the length of stem, roots, number of leaves, fresh growth of flowers, *etc.*

Development

It refers to the qualitative changes from the initiation of growth to the death of a plant or plant parts like seed germination, formation of flowers, fruits and seeds, emergence or buds, leaves and falling of leaves.

Plant Growth Regulators

Plant growth regulators are organic compounds other than nutrients, which in small amounts promotes, inhibits or modify the physiological processes in plants.

Phytohormones

The phytohormones are the organic substances that are produced naturally in one part and usually translocated to other part of plant. Where in the very small quantity it influences the growth and other physiological function of the plant. The term growth regulator is used for the substances which work similar to the phytohormone but synthetic in the nature.

Hormones

It may be defined as substance synthesized in one part of the plant and translocated to another part, where they exert their effect *i.e.,* the site of production and site of action are different. Phillips (1971) defined growth hormone as substances which are synthesized in particular cells and transferred to other cells where in extremely small quantity influence development process. There are different categories of plant growth substances, which may be broadly classified into growth promoting and growth retarding substances or into naturally occurring growth substances and synthetic growth substances. However, nowadays the term growth regulating substance is preferred because it include both naturally and synthetic growth substances. Growth regulating substances may be classified into following categories.

1. Auxin-Precursor Tryptophane

Auxin are growth regulators, which at low concentration (<.001m) promotes growth along longitudinal axis of shoots.

- Indole auxins - IAA, IBA, IPA
- Nepthyle groups- NAA, NOA
- Phenoxy group - 2,4-D; 2-4,5-T
- Benzoic group 2,4,6- Trichlorabenzoic acid, 2,3,6-Trichlobenzoic acid

2. Gibberellins-Precursor Terpenoids

Gibberellins are hormones, which promotes plant growth specially of stem. It is found in Gibberellic acid – GA_1- $GA_{120.}$ Young leaves are the site of active GA synthesis. GA synthesis also occurs in root, embryo and endosperm.

- It promotes seed germination.
- Cell elongation in intact plants.
- Help in breaking rest and dormancy.
- Prevent flower initiation.
- Induce parthenocarpic fruit set.

3. Cytokinin-Precursor 5-AMP (Isopentenyl Group)

The cytokinins are substances, which primarily act on cell division and have little or no effect on extension growth, *e.g.,* Benzyle adenine, Kinetin, Zeatin, *etc.*

- Initiation of cell division
- Delaying senescence
- Enlargement of cells
- Differentiation of cells (interact with auxins)
- Induces the flowering in short-day plants

4. Abscisic Acid–Precursor Sesquiterpenoids Pathway (Mevalonic Acid)

- ☆ Bud dormancy
- ☆ Stimulates the closure of stomata
- ☆ Induction and maintenance of dormancy
- ☆ Disease resistance
- ☆ Protecting cells from dehydration

5. Ethylene-Precursor of Methionine

Gases present in smoke could modify plant growth. Neljubor (1901) was the first to show the importance of ethylene present in illuminating gas as a growth regulator of plants. Virtually all living plant cells are capable of producing it. An unique feature of this plant hormone is that, it is regularly flushed out of the plant.

- ☆ It promotes germination of seeds.
- ☆ Seed dormancy is overcome.
- ☆ At higher concentration growth of the sprout is checked.
- ☆ It has enhances the ripening of fruits.
- ☆ Accelerate the latex flow from rubber plant.

6. Inhibitor

- ☆ It may be natural as well as synthetic, *e.g.,* ABA, MH, Dormin, SADH.
- ☆ It supresses the growth of apical meristem.
- ☆ Higher concentration of inhibitors may cause malformation.

7. Retardents

- ☆ Retardents only slow down the growth of the apical meristem.
- ☆ Higher concentration of retardants does not cause any malformation to the plant, *e.g.,* CCC. Phosphon- D, AMO-1618, Alar, Phos-S.

8. Flowering Hormones

Florigen, Anthesin

Selection and Using Plant Growth Regulators in Floriculture Crop

Plant growth regulators and chemicals that are design to affect plant growth and development are applied for specific purpose to affect specific plants response.

Optimizing Result

For best result plant growth regulators should be handled as production tools like water and fertilizer. They are most effective applied appreciable time to regulate plant growth and development. In other word growth retardant cannot shrinked over grown plant.

1. They reduce the growth rate of plant improve its colour and general condition, *e.g.,* Cycocil and B- nine.
2. They increase plant branching for enhanced cutting production or for a more bushy potted plant or hanging basket flower, *e.g.,* Florel.
3. They enhance flower initiation or shrinkonized flower, *e.g.,* Cycocel florel.

Role of PGRs in Floriculture

The use of plant growth regulators in modern horticulture is well stabilized more particularly horticulture industries in advance countries and their application is fast picking up in the developing nation as well. Thus their application in these crops has been extended to alter the variety of responses. Some plants varieties or cultivars are responsive to specific plant growth regulator but not all plant growth regulators. The hormones are requiring to minute quantities to produce large specific physiological process. Ornamental crops find extensive use growth regulator for modifying their developmental process with in the broad groups of plant hormone. Some act as a growth promoter and other act as growth retardant.

The major areas were growth regulator are used in floriculture are –

- ☆ Propagation.
- ☆ Plant height control.
- ☆ Breaking dormancy.
- ☆ Promotion of vegetative growth.
- ☆ Regulation of flowering.
- ☆ Improvement of spike and flower quality.
- ☆ Increasing the yield of flower, bulb corm.
- ☆ Extending the Vase life of flower.

1. Plant Propagation

There are 3 method of propagation employed in flowers viz. asexual or vegetative method, sexual (through seed) and micro propagation. Majority of ornamental plants are propagated by vegetative method as stem or leaf cutting, bulb, corm and tubers.

a. Vegetative Propagation

Rose, Carnation, Chrysanthemum and Gerbera, Auxin (IBA, IAA, NAA) or extensive use for rooting of the cutting the most common and widely used auxin in promotion rooting in IBA followed by NAA and IAA.

Usage of Auxin

- ☆ Quick deep method deeping the basal portion of cutting in their concentration of 1000 to 10000 ppm (1ppm=1ml or 1 per cent =1000ppm) for 5sec to 2 minute depending upon the nature of the cutting whether they are soft or hard wood cutting.
- ☆ Deeping the wet portion of cutting in mixed with auxin (500- 12000 ppm).
- ☆ The success of rooting depend upon the some external and internal factor like rainy season is the best photoperiod, light, temperature, aeration, humidity and nutrient status of the cutting, *e.g.,* - For Bougainvillea hard wood cutting are use for their propagation 1000-3000 ppm, IPA is quick deep method is best, NAA at 1000 ppm increases the percentage of rooting and reduces the time taken for rooting.

b. Sexual or Seed Propagation

All seasonal flower Marigold, Pansy, Petunia, Sweet alyssum, Lupine, *etc.,* are propagated by seed. Seeds are treated with lower dosage of GA which increase germination per cent seedling bigour and final population stand, *etc.*

c. Micro Propagation

In Orchids, Anthurium, Carnation and Gerbera, cytokines and auxins are widely used and supplements for induction of shoots and roots respectively in the tissue culture plantlets.

2. Plant Height Control

Growth retardant like Maleic Hydrazide (MH), Cycocil (CCC) and B-nine are used for height control of ornamental crop specially those grown in pots and terraces. An ideal growth retardant should be one which will be universally effective non-phototoxic and will prolong the post-harvest life without living any harmful residual effect. Many growth retardant are available commercially greenhouse production.

3. Breaking Dormancy

The chemical like Ethylene, Chlorohydrins (Ethera), Ethephon, Thiourea and Hydrogen cynamide (Dormex) are used for breaking dormancy of gladiolus corm and cormels. Soaking in 1000 ppm ethereal for 24 hours breaks the dormancy in gladiolus. Plant growth regulators like propilgallade and MH can be used for extending the dormancy period.

4. Promotion of Vegetative Growth

Stem length is an important quality criteria in many commercial flower crops foliar spray of GA_3 100-400 ppm can be applied to achieve this like Rose, Chrysanthemum, Carnation and Dahlia.

5. Regulation of Flowering

GA_3 at 5-25 ppm causes early flowering in antirrhinum, GA_3 10 ppm applied 140 days after planting increase number and size of flower. GA_3 100 or 200 ppm, in Dahlia induced early flowering. In general IAA and NAA at higher dosage more than 100 ppm for delay flowering 2, 4-D 1 ppm and IAA 150 ppm at early bud stage increases the flower size.

6. Increasing the Yield of Flower

GA_3 at lower concentration 5-10 ppm improve flower set and seed yield in many flowers, *e.g.,* Pansy, Petunia and Phlox, *etc.* Kinetin, BA 100 ppm, GA_3 100 or 200 ppm once or twice as foliar spray on flower the corm and bulb have good yield in gladiolus and tuberose flower.

7. Extending the Vase Life of Flower

The vase life of cut flower can be prolonged significantly by adding kinetin HQC (Hydroxy quinoliae citrate), $AgNO_3$ (Silver nitrate), STS (Silverthio-sulphate), Citric acid and Sucrose compound in the vase water, *e.g.,* Rose, Chrysanthemum, Tuberose, Gladiolus and Gerbera.

Chapter 17

Styles of Garden

Styles of garden is a good appearance typically determine by the principle according to which gardening can be worked out. Besides the term landscape gardening the other two familiar terms in gardening are the formal and informal gardens. The following terms are given below:

1. **Formal:** In this garden the design is stiff as everything is done in a strait and narrow way. In such garden everything is planted in strait lines, *e.g.*, Mughal, Persian, French and Italian.
2. **Informal:** In an informal garden, the whole design is informal as the plants and the features are arranged in a natural way without following any hard and fast rules, *e.g.*, English, Japanese and Chinese.
3. **Free style:** It is study of the both formal and informal garden which is found middle of the systematic and non-systematic in nature, *e.g.*, Rose garden (Chandigarh).
4. **Wild:** The concept of wild garden is not only against all formalism but it also breaks the rule of landscape styles and naturalize plants in shrubberies. Wild style garden are found unsystematic planting of trees, shrubs and bulbous plants, *etc.*
5. **Ornamental garden:** In planning a garden several factor like size of the house and the space available for garden, availability of border, cost of laying the garden and its maintenances have to be taken into consideration, *e.g.*, Park (Private and public).

Though in India from history and ancient literature we find that gardening was quite vague in older times, but unfortunately there is no garden style called 'Indian Garden' which can claim a place in the major gardening styles of the world.

The famous garden styles of India the "Mughal gardens" are nothing but a replica of the ancient Persian gardens.

The features of the major garden styles are given below:

(A) Feature of Formal Garden

- ☆ First plan is made on paper and then land is selected accordingly.
- ☆ Land is leveled.
- ☆ Symmetrical design.
- ☆ Geometrical (Square, Rectangular, Circular beds border).
- ☆ Roads and path cut a right angle.
- ☆ Balance is symmetrical as a same feature replicated on both side of center axis.
- ☆ Hedge, edge, topiary trimmed.
- ☆ Tree can be selected as individual feature.
- ☆ They are usually small to medium in size.
- ☆ It has East and West orientation, *e.g.,* Mughal garden, Persian, Italian, French and American.

(B) Feature of Informal Garden

- ☆ Plan is forced to fit the land.
- ☆ Main aim is to capture natural scenery.
- ☆ Land is not leveled.
- ☆ Asymmetrical design.
- ☆ Non-geometrical beds and border, untrimmed hedge, edge, and topiary.
- ☆ Individual plant as not selected as feature.
- ☆ Much more variety of elements is used.
- ☆ Animal like, umbrella, seats, cassettes, rockery, water pull, *etc.* Such as Japanese, Chinese, English.

(C) Feature of Free Style Garden

- ☆ A new approach to gardening that allows developing a garden with what on hand.
- ☆ This style combine the god point of both formal and informal as well as naturalistic feature and aesthetic mixed to create a picturesque effect.
- ☆ This style is suited to almost all situations.
- ☆ This style garden are found as Rose garden.

(D) Feature of Wild Style Garden

- ☆ Wild styles garden no rules are followed but aim is to make the garden beautiful and natural.
- ☆ William Robinson gave the concept of wild style of gardening.
- ☆ To make garden more beautiful and natural.
- ☆ Such gardens are laid out for more agreeable communication in nature.
- ☆ Wild variety of trees, shrubs and creepers are used in natural ways.
- ☆ No formal rules are followed and allowed to give in their natural shapes.

Chapter 18

Types of Garden

The major garden types which shall be discussed in the following pages are: English garden, Mughal garden, Persian garden, Italian garden, French garden, Japanese garden, *etc.* Out of these the Mughal, Persial, Italian and French styles fall in the category of formal gardens, whereas the English and Japanese gardens are classified in the informal style of gardening.

The famous features of the major types of garden are given below:

(A) Special Feature of English Garden

- ☆ The famous British garden architects Repton and Capability Brown advocated the concept that the British gardens should look like the countryside.
- ☆ In it lawn has mixed border especially of herbaceous annuals as well as herbaceous perennials, shrubbery and rock garden.
- ☆ The English climate suits admirably well for the growth of herbaceous annuals.
- ☆ Typical grassland climate in England.
- ☆ Gardening is hobby of rich people.
- ☆ English men were very fond of flower essential feature. (Lawn, Rockery, Herbaceous border).

(B) Special Feature of Mughal Garden

- ☆ Site near hill, slope with perennial rivulet.
- ☆ Gardens are enclosed with walls and fitted with tall gate.

- ☆ Garden has at least 7, 8 or 12 terraces symbolizing 7 planets, 8 paradises or 12 zodiac with entrance at the lowest terraces.
- ☆ Running water in canals.
- ☆ Terminal building.
- ☆ Baradari with 12 doors, 3 in every direction (Safed baradari in Lucknow).
- ☆ Symbolism and plant material.

(C) Special Feature of Persian Garden

- ☆ Based on idea of heaven.
- ☆ Strictly formal and symmetrical.
- ☆ Beautiful architectural work.
- ☆ Laid out after cutting terraces.
- ☆ Water flowing canals.
- ☆ Plants cypress a symbol of eternity.

(D) Special Feature of French Garden

- ☆ Formal garden in perfection.
- ☆ The moral of French garden style of Le Notre seems to teach the lesson 'how to think big'.
- ☆ Unexampled scale of mass and sweep of design.
- ☆ This style dominated the gardens of civilized Europe for a long time.

(E) Special Feature of Italian Garden

- ☆ Came into existent at the Renaissance time.
- ☆ Resemble Mughal or Persian garden.
- ☆ Fountains, sculptures, water canal, box on yellow, hedge, topiary arbaur, trellis and architecturally beautiful garden, *e.g.,* Plant of rose.

(F) Special Feature of Japanese Garden

- ☆ The both the Persian and Japanese garden's design were based on their respective ideas of heaven.
- ☆ The Japanese continued the same style for centuries but still remain popular.
- ☆ Ornamental water ponds, streams waterfall, fountains, wall, water basin.
- ☆ It should be a placed where the mind finds rest and relaxation.

Plants in Japanese Garden

- **Evergreen** – Abies, Chytomeria, Juniperus, Mangolia, Podocarpus, *etc.*
- **Decidous** – Acer, Popolus, Morus, Salix, Prunus, *etc.*
- **Shrubs** – Azalea, Gardenia, Callellia, Lagertroemia, Rhododendron, *etc.*
- **Climber**–Clematis, Wisteria, Loocera, *etc.*
- **Annuals** – Aster, Chrysathemum, Coronation, *etc.*
- **Bulbous**- Canna, Gladiolus and Lily, *etc.*

Types of Japaneese Garden

A Japanese garden may either be in the form of a large public park or a small family garden designed for living which is seen usually by members of the family or he family guests. The Japanese gardens are further classified based on positions, shape, and purpose. The important types are: i) hill garden, ii) flat garden, iii) tea garden, iv) passage garden, and v) sand gardens.

1. Hill Garden

Out of the various styles, this one is considered to be the ideal by many garden enthusiasts. This style is known in Japanese as Tsukiyama-niwa or Tsukiyama sansui, meaning hills and water. Though this garden can be laid out in a space of any size by making use of perspective to maintain a reduced scale, it is obvious that a fairly large area is much suitable as laying out a mountain scenery needs considerable space. The hill garden is made up of one or more hills designed with earth mounds and exposed weathered stones. The other features of this garden are water in the form of a stream or a pond or a waterfalls or all the three with or without islands and also bridges, lantern, stones, and trees. If it is not possible to introduce water in the garden, its presence is effected by a dried bed of stream or a dry shoreline. The important points in the garden are decorated with stones and selected trees. But pine trees may be planted to give the effect of being swept by wind. Untrimmed stepping stones are placed over the walks.

An island is general an usual feature in a hill garden. When the island is present it should be decorated with a "Worshipping stone", called raithai-seki in Japanese, a "Snow-viewing" lantern and a pine tree.

2. Flat Garden

As the name implies, Hira-niwa or flat gardens are laid out in flat ground without hills or ponds. Flat gardens are supposed to represent a mountain valley or a meadowland. These gardens were popular during the era of Muromachi (1392-1573).

A Flat garden is not necessarily as flat as a pan-cake. Since it stimulates a mountain valley, low rounded hills designed with the help of stones or earth mounds or both will look quite appropriate in a flat garden. But some others opine that there should not be any ups and downs in a flat garden. The usual features to break the monotony of a flat garden are a well, a water-basin made of stone in the shape of an urn, stones lying close to the ground, stepping stones, and trees. The trees are trained to lie close to the ground. In a flat garden, the principle is to avoid strong vertical lines represented by tall pines. In olden days the flat gardens were simple in design consisting of a few low-growing trees and flat rocks but in modern times many features of the tea garden, like water-basin, have been incorporated in this type of garden also.

3. Tea Garden

The tea garden is laid out based on certain principles and customs of the Japanese tea ceremony and hence needs a considerable space of at least about square metres, for its designing. Since the performance of the tea ceremony needs an atmosphere of intimacy it is essential that the garden be enclosed by a fence. But the fence should be rustic in nature, with a gate made of very light material such as bamboo. To protect the tea house from the noise of the outer world, the tea gardens are divided into an outer garden (soto-roji) and inner garden (uchi-roji).

The outer garden is comparatively a narrow area, with a waiting place where the guests are supposed to wait until the master of the house appears to welcome them. This waiting place has a water-basin for the convenience of the guests who can wash their hands and a stone lantern for illumination, but in present days this serves more as a decoration piece. A stone path, usually of stepping-stones, leads to the inner garden. The inner garden is also separated from the outer garden by a rustic fence and a gate made of light material.

4. Passage Garden

The passage gardens, the Roji-niwa, are those which are laid in narrow passage, as for example a narrow space between two houses or approaches to buildings. As such areas are generally narrow the garden lay-out should be simple and not overcrowded.

In such gardens there should be hardly any ornaments such as lanterns, basins or other man-made features. The common features of a passage garden are a few key rocks, slabs of stones, and only a couple of types of plant. Bushy shrubs and trees are unsuitable in a passage garden; instead, plants with open form and slender shapes are selected.

5. Sand Garden

It is the simplest style of gardening, though not liked by many as it is totally devoid of plants. The most famous sand garden exists in Kyoto and is known as Ryoanji garden. The garden consists of a rectangular area of about 350 square metres adjoining a Zen Buddhist temple. The main feature of this style of gardening is to arrange a few vertical and prostrate stones in groups of 2 or 3 and to fill in the gap between the stones with fine white gravel. The gravel is raked in most simple patterns simulating the ripples flowing water. The raking has to be repeated often to keep the garden in its best shape. This style of garden looks pleasant and effective only when confined to a limited area.

Chapter 19

Flower Show and Arrangement

Flower Show

Flower show is an event at which flowers are displayed and judged for awards. Flower show help to spread the world the about the joy of gardening and serve as forum for learning and discussing, the latest horticultural design and trends. In addition many national level flower shows offers training opportunity for members who are interested in becoming certified flower show judged. Flower show is success depends upon the co-operation between exhibitor and members of the flower show committee.

Important Feature of Flower Show

- ☆ It provides knowledge about the wide range of plant that can be grown in locality.
- ☆ It provides a chance to see all the best materials at a time in one place.
- ☆ It is also an occasion to demonstrate how a perfect exhibit can be grown.
- ☆ It has an aesthetic value.
- ☆ A stage of discussions and solutions related to growing of plants.

Benefits of Flower Show

1. The planning and staging of flower show develops affinity and togetherness among the people.
2. Provide a chance to meat and connect with potential workers.
3. Conducting the show to educates the people in floral design and gardening.

Preparation of Flower Show

1. A general or a ground committee should preparation the general arrangement.
2. Such a selection of site arranging tent samiyanas and chairs, *etc.*
3. An editorials committee should prepare the schedule well in advance prepare rules and regulation, printing admission ticket under take necessary.
4. The floral committee will look after the cut flower and floral arrangement.
5. The finance committee whose duty will be to look after the different plant sections.

Rules for Flowers Exhibition

- ☆ A small amount of fee is generally charged.
- ☆ An entry form should be filled in by the applicant.
- ☆ The competitors may inter more than one exhibit in a section but will be awarded only one prize.
- ☆ Arrangement of potted plant must be complete within the given stipulated time.
- ☆ The entry should be correctly and legibly leveled for bonus point.
- ☆ The exhibit should be accompanied by an entry card.
- ☆ Exhibitors name should not appear on any exhibit before judging commences.
- ☆ Pot plant will be judged on the basis of number, size and quality of blooms.
- ☆ Cut flower will be judged based on their size, form and quality.
- ☆ Division of large garden into small or medium will not be accepted.
- ☆ All exhibits in cut flower section must be from plant which has been grown by the exhibitors for not less than 3 month period to the show.
- ☆ All cut flower to be exhibited on their own stem which should be long enough to show the blooms at least 5cm above the rim of the container.
- ☆ No person shall be allowed to be a judged in a section in which he or she is a competitor.
- ☆ The running challenge shield, cups, *etc.* may be retained by the winner and must be return undamaged at least one month before the date of next show.

Role of a Flower Show Convener

1. **Goal of flower show-** To educate society members and the viewing public in floral design, stimulate interest in horticulture and provide an outlet

for creative expression. Members see new cultivars and learn what trees, shrubs, flowers and plants grow locally.

2. **Qualification of flower show convener-** For a show convener to be able to produce a good flower show, they need to have an extensive knowledge of the variety, plant species, cultivar and common and Latin words. The convener should have interned flower design and horticultural show both in and outside the society.
3. **Scheduling-** The show convener write a yearly show schedule in to division horticulture and design. The schedule lost specific wording to include all essential details of the show. A standard flower show has a minimum of 5 classes or category in each division with a minimum of and exhibit in each class. The schedules need to confirm to the season selective flowers, fruits and vegetables, *i.e.,* expected to be in peak condition at the time of show. Once the dates, classes and themes for design have been set the schedule is printed and made available to all members.
4. **Staging-** Staging is done by the show convener with the flower show clerk, tables must be set up at a standard height with neutral colourful clothes as well as signage to identify division, dividends between classes and correctly size inches for standard and miniature design extra lighting may be provided for judging. Flower show entry tags must be made available to exhibition both prior to as well as at the show.
5. **Placement and classification-** The show convener helps the exhibitors fill in entry tags and with the placement of their specimen for cultural or design classes if needed. Prior to the judge arriving the exhibit area is closed to member to allow the show convener to check for the correct placement of exhibit moving those that may have being placed in the wrong section and removing those that do not confine to the schedule, *i.e.,* the wrong types of flower show conveners may help members in the grooming of dead foliage are below prior to the show when necessary. If a class attracts a large number of entries the show convener may sub-divide the class. There must be minimum of three specimens in each sub-division.
6. **Judges-** The show convener will select and invite a judge well in advance of the show. Convener must provide the judge with the address, show schedule and rule of the show prior to the show. The convener has a list of accredited judge and their location to choose from. Only the show convener and clerk may be present by judging takes place. The convener accompanies the judge to answer any further question that may arise during judging.
7. **Award-** Annual award for design most points and novice of the year and give to member at the annual general meeting and are presented by show convener.

8. **Education-** A variety of workshop covering a wide range of topic is offered to member workshop is given by judge or convener. When a judge is presenting a workshop the convener is responsible for the logistics for the event, *i.e.,* arranging for the presenter to send, booking the venue, ensuring material are provided helpful to society member and collecting any fee if applicable.

Different Categories or Classes of Flower Show

Some most important classes are given below-

- **Class 1**- Which are open for all exhibitors to display the flowering, foliage and pot plants, *e.g.,* Annuals, cacti, ferns, *etc.*
- **Class 2**- In which the display of the all pot plants generally hight is 30cm. the flower show open for all exhibitors in one colour or mixed colour, *e.g.,* Clendula, geranium, phlox, petunia, *etc.*
- **Class 3-** It is open for all exhibitors. In which the use of hanging basket of flowering and foliage plants, *e.g.,* Annuals, perennials, succulents, ferns, bulbous plant, *etc.*
- **Class 4-** It is open for amateurs only. In which the use of pot plants (upto 30cm).
- **Class 5-** It is open for exhibitors. In which the use of pot plants, *e.g.,* Bulbous plant.
- **Class 6-** It is open for all exhibitors. In which the use of cut roses, pot roses, greenhouse roses, new varieties of roses, *etc.*
- **Class 7-** It is open for all in which the use of cut flowers other than roses, *e.g.,* Tuberose, gladiolus, gerbera, orchids, bird of paradise, chrysanthemum, carnation, dahlia, *etc.*
- **Class 8-** It is open for all, in which the use of arrangement of flower/ foliage. It is divided in different section viz. vase/bowl- eastern style, western style, beautiful arrangement (rangoli), flowers, petals, *etc.* garland, corsage and floral craft.
- **Class 9-** It is open for arrangement of flowers for gardeners (Malis). There is no entry fee. In which the use of mixed flower and foliage, bouquets, garland, wreath, floral craft, *etc.*
- **Class 10-** It includes all type of gardens.
- **Class 11-** It is used for landscape models. It is divided in various section, *i.e.,* formal style, informal style, free style, Japanese style, English style, arches, pergola, topiary, *etc.*

Flower Arrangement

Flower arrangement may be defined as the art of organizing and grouping together flower and plant material to achieve harmony of form, colour and texture. In other word arranging flower is a kind of art and a successful flower arranger is a borne artist, but one can become regional conversant about this art with proper training a lot of patience and perseverance.

Types of Flower Arrangement

1. Japanese or Oriental or Eastern Type

It means Japanese flower arrangement is also called "Ikebana" (Kado). Emphasis is given on spiritual and religious background and only few flowers are used. "Ikenabo" was the first school of "Ikebana" meaning the arrangement of flower of the hermitage along a pond which was started by "Buddhist Monk Sammu" around A.D. 621.

2. English or Western or European Type

Emphasis given on mass flower arrangement and is primarily a form of art so as make the arrangement attractive as for as possible.

Type of Ikebana

- ☆ Moribana
- ☆ Nageire
- ☆ Jiyubana
- ☆ Zeneika
- ☆ Zeneibana
- ☆ Morimona
- ☆ Tatebana

1. Moribana

- ☆ It means piled flower.
- ☆ Arrangements are made in sallow container.
- ☆ Flower, branches supported pin holder.
- ☆ Look very natural.

2. Nageire

- ☆ Arrangement of upright container of tall bases.
- ☆ Flowers have sufficient stem length.
- ☆ Supported with the help of cross bark fixture.

3. *Jiyubana*

- ☆ It can be arrangement both moribana or nageire style.
- ☆ Wood, metal any other materials may be used in addition to flowers.

4. *Zeneika*

- ☆ Straight material with uneven height is used.
- ☆ This style dose not simulate in nature.

5. *Zeneibana*

- ☆ In this style a beautiful sculpture is created using wood, stone, rocks, metal depicting any natural scenery.

6. *Morimona*

- ☆ In this style fruit, vegetables and flower are arranged.
- ☆ The fruit and vegetable should be chosen and grouped for variety of texture, colour and shade.

7. *Tatebana*

- ☆ It is arranged, mental and religious.
- ☆ Plants with branches are twigs placed upright in the vase.
- ☆ Bottom grasses are placed around the court.

Material Requires for Japanese Flower Arrangement

- ☆ Flower vase
- ☆ Holder
- ☆ Fixture (single, split and cross bar)
- ☆ Flowers
- ☆ Foliage
- ☆ Fruits or vegetables
- ☆ Wood

Principle of Flower Arrangement

- ☆ Hormony
- ☆ Simplicity
- ☆ Variety
- ☆ Balance (symmetrical and non-symmetrical)
- ☆ Proportion
- ☆ Emphasis or ascent or focal point
- ☆ Space

English Flower Arrangement

It is a decorative display of cut flowers where emphasis not given on individual plant material, here the arrangement give a mass effect and the deigns looks more attractive. It gives a flowering, radiating effect.

Type of English Flower Arrangement

- ☆ Triangular shape.
- ☆ Circular shape.
- ☆ Crescent shape.
- ☆ Fan shape.
- ☆ Hogarth or S-shape.

1. *Triangular Shape*

- ☆ Use in personal and professional purpose.
- ☆ Height and width is fixed with flower.
- ☆ Focal point is stabilized.
- ☆ Mostly display on buffet table or in the side stand.

2. *Circular Shape*

- ☆ Adds a pleasing element.
- ☆ Satisfy the viewer eye.
- ☆ Lay on conference table or on buffet table.
- ☆ Can be presented to high class executives or political occasion.

3. *Crescent Shape*

- ☆ Also known as half-moon shaped arrangement.
- ☆ It is asymmetrical and formal.
- ☆ It is very eye catching.
- ☆ Kept in lobby of hotel.
- ☆ Used as a focal point to catch the attention of the guest.

4. *Fan Shape*

- ☆ It is a low arrangement.
- ☆ Generally placed in the restaurant either in buffet or on dining table.

5. *Hogarth or S-shape*

- ☆ This is very graceful style of arrangement.
- ☆ It is easy to make when curved branches are used.
- ☆ Flowers or field on the centre of "S"-shape arrangement.

Difference between Eastern and Western Arrangement

- ☆ Primarily eastern style is asymmetrical but western style is symmetrical style.
- ☆ Eastern style impress more by the beauty of individual material, but western style employ more flowers to create mass effect.
- ☆ Accessories are never used in western arrangement, but in 'Ikebana' interesting branches, drift could pieces of bark, shells, *etc.* are used the initiate the nature.
- ☆ Contrary to arrangement of material in Western style in Eastern style never touch the rim of the base.

Indian or other Type Flower Arrangement

1. **Garland-** Floral garlands are used for offering to gods, for weddings, for decoration, *etc.* Garlands can be made of different type of loose flower, *e.g.,* Marigold, Rose and Jasmine, *etc.*
2. **Gajra or veni-** These are the floral ornaments used to decorate the hairs by Indian women. Gajra are generally worn by north Indian women that usually consist of Chrysanthemum, Tuberose and Rose flowers along with foliage of thuja. Whereas, veni is common among south Indian women. A veni can be made up of flowers of Crossandra, Tuberose and buds of Jasmine.
3. **Rangoli-** It is the decoration of colourful flowers in different design in association with colourful sand. Flowers and petals of Marigold, Rose, Chrysanthemum and foliage of *Murraya exotica*, Thuja and Fern, *etc.*
4. **Bouquet-** These are used during welcoming and departure of guests and relatives. Generally conical, round or flat types of bouquets are made. In flat type of bouquets generally spikes of Gladiolus, Tuberose in combination with Gerbera flowers are used.
5. **Wreath-** It is an arrangement of flowers, leaves, or stems fastened in a ring and used for decoration or for lying on a grave. Generally used in spring flowers.

Chapter 20

Protected Cultivation

Greenhouse

A greenhouse is a specially constructed structure for growing plants under control condition. The greenhouse structure has no green colour but its name perhaps derived from the fact that plant (green) are grown in it. Greenhouse is also known as glass house and poly house. In United States a greenhouse refers to a structure covered with a transparent material that transmits light from plant growth whereas, in Europe a greenhouse is an artificially heated structure for overcoming climatic adversity while making use of solar-radiation. In Indian climatic conditions the term greenhouse includes a glass house for plants and any other structure including lath house for growing shade loving greenhouse plant. Greenhouse protect the crops against cold, rain, trails and wind for provided plant with improved greenhouse crops can be produced out of the season, year round, their yield and quality of higher than these growth in open field.

Greenhouse Effects

Greenhouses covering materials (plastic, glass or fiber glass) allows the solar radiation to pass through it but traps the thermal radiation emitted by the plants inside which in turn increase the temperature of greenhouse. This is also known as a greenhouse effects.

Location for Greenhouse

Greenhouse should be located on a slightly raised ground and should preferably not be situated along the boundary as large trees of the neighbor growing along the boundary may interfere with the greenhouse life. The size of greenhouse will depend upon the space available, the financial capacity and the interest of individual growers. In the mild morning the sunlight is preferred and the afternoon sun is excluded by some lattice work of bamboo or wood or growing creepers on the western side's location site of greenhouse in north and south for best production of the plant.

Advantage of Greenhouse Effect

- ☆ Protection of flower crops from adverse weather and climatic conditions.
- ☆ Production higher quality and quantity of flower round the year with good export potentiality.
- ☆ Higher net return per unit area protecting from biotic and abiotic stresses.
- ☆ Efficient use of land, water, temperature space and natural resources such as sun light, temperature, relative humidity and atmospheric CO_2.
- ☆ Easy management of insect, pest, diseases and weeds.

Types of Greenhouses

There are 3 basic types of greenhouses, attached, detached (free standing) and connected. These types of greenhouse are constructed in several 'T' style such as

- ☆ Even span (systematic).
- ☆ Uneven span (unsystematic).
- ☆ Quonset (cemented).

Greenhouse Environment

The greenhouse environment can be divided into two distinguished parts –the areal environment and the root environment. The physical factor namely light, temperature, humidity, wind, velocity and CO_2 concentration that need to be controlled in the area environment are the often referred as the greenhouse climate.

1. **Temperature-** The temperature of greenhouse varies from about 10-25°C for the almost all flower crops.
2. **Heating system–** Greenhouse can be heated through fuels (natural gases, fuel oil and coal), hot water stream infrared radiation, heater, solar radiation system, steamed temperature 100-107.7°C from a central boiler through network of pipe.
3. **Light-** On the clear summer day plants in greenhouse may receive light as much as 120000 feet candle/122000 Lux of light which is an excessive amount for most of the crops. Most of the greenhouse plants required light 3000-10000 feet candle/3280- 10800 Lux.
4. **Humidity-** Humidity is an important factor in greenhouse production. Humidity between 90 per cent and 55 per cent respectively does not affect the growth and development of greenhouse crop.

Low Cost Poly House

Low cost poly houses are small structure that provide temporary crop protection. These are also called miniature greenhouse. The low cost poly house is made of polythene sheet of 700-800 gauges (1 gauge/0.25 micron or 0.00025 mm). This is supported by the bamboo on GI pipe which has diameter 15 mm.

Low Tunnels

Plastic low tunnels are flexible transparent covering that are installed over rows of individuals beds in open field during winter. Low tunnels enhance plant growth by warming the air around the plant (micro climate). They can also warm soil and protect plants from natural disaster like hells injury and cold wind. Low tunnels of convenient size 0.75- 1m width are made using UV stabilized "LDPE" film corrugated on plane fiber rain forced plastic sheet.

Principle of Greenhouse Cultivation

- ☆ The greenhouse is covered with a transparent material such as plastic, PVC sheet or poly carbonate sheet or glass or fiber or micro plastic.
- ☆ Based upon its transparency the greenhouse cover transmits most of the sunlight.
- ☆ The crop floor and other object inside the greenhouse absorb the sunlight admitted inside the greenhouse .
- ☆ These objects in turn emit long wave thermal radiation for which the greenhouse covering material has lower transparency. As a result of these the solar energy is trapped thus leading to increased temperature inside the greenhouse .

Indo-American hybrids seeds company has made greenhouse for cultivation of ornamental plants in 1970. More than 250 companies grown an ornamental plants in India.

Advantage of Growing Flower Crop in Greenhouse

1. Insures the production of any plant at any place and throughout the year.
2. Easy to control insect pest and disease.
3. Water requirement reduces.
4. Labour requirement is less.
5. Earliness as it reduces crop duration.
6. Ideal temperature.
7. Amendment sunshine throughout the year special in rainy season autumn and winter season.
8. Shorter production cycle.
9. Highly production.

Basic Concentration for Flower Cultivation in Greenhouse

- Feasibility.
- Type of greenhouse structure.
- Planting material.
- Growing system (Ground beds, raised beds, bench and pot).
- Plant protection.
- Post-harvest handling.
- Management and coordination.

Important Flower Growth in Greenhouse

1. Cut Flowers

Rose, Carnation, Jasmine, Chrysanthemum, Lilium, Gerbera, Anthurium, Tulip, Orchids and Statice, *etc.*

2. Loose Flowers

Jasmine, Chameli, Marigold, Gladiolus, Bird of Paradise, *etc.*

Chapter 21

Civic Aspects of Ornamental Horticulture

PARK

Park is an area of natural, semi-natural or planted space set aside for human enjoyment or civic aspects and creation for the protection of wildlife or natural habitat. It may consist of grassy areas, rocks, soils, trees as well as building and other artifacts such as monuments, fountains or play ground structure. Many parks have traits for walking, biking and other uses for civilians. Some parks are built adjacent to water bodies or water courses and may comprise a beach or boat dock area.

The big parks can be natural areas of 100-1000 square km with abandon wildlife and natural feature such as mountains and reverse. There are also amusement park or the larger type of park in the world. In many larger park camping in tents is allowed with a permits. Large national and sub-national park are typically overseen by park ranger and park warden.

Park Design

Park design is influenced by the intended purpose and audience as well as by the available land feature. A park intended to provide recreation for children may include a play ground. A park primarily intended for the adult may feature walking paths and decorative landscaping specific feature such as riding trails may be included to support specific activities. Different landscaping and infrastructure may even affects children rates of the use of park. A park is a part of urban infrastructure for physical activities for families and communities together and for socialization.

Study reveals that people who exercise outdoor in green space derive better mental physical health benefits. These activities for all age and income group are important for physical and mental well being.

Step by Step Plan for Creating Park

- Secure the community commitment.
- Convene steering committee.
- Choose a site.
- Plan.
- Identify and secure potential partner.
- Secured long and short-term funding.
- Schedule work days in advance.
- Plan a big work day.
- Implement a maintenance plan.
- Pursue consistent engagement.

Legends of the Park (Model)

- Entry point of vehicle into the shared zone.
- Seating arrangement around the green and the shared zone.
- Garden area with seating.
- Possible location for toilet.
- Rain garden (to be completed by adjoining development site).
- Picnic area (to be completed by adjoining development site).
- Seating walls and sets as to the green.
- Open plaza area for events with a pergola for shade and seating.
- Main East and West path.
- Play area.
- Seating area.
- Car parking area.
- Potential area for a community garden.
- Picnic and social area.
- Feature tree for shade and visual distinction.
- Exit point for vehicle.
- Proposed building.

Role in City Revitalization

City parks play a vital role in improving cities' features for residence and visitors.

Design for Safety

Parks need to be safe for people to use them. Research shows that perception of safety can be more significant in influencing human behaviours. If citizens find a park unsafe then they might not make use of it at all.

Types of Park

1. Private Parks

Private parks are owned by businesses individual and are used at the desecration of the owner. There are few type of private parks and some of them are privately maintained and used for public aspect.

2. Mini Parks

They provide quality park and recreation space for inner city residence and challenged by the limited amount of available park space in urban areas. As a result of diminishing access to park and open space to physical recreational needs of the urban youth. To meet these growing need parks and recreation agency are trying to play an important role conversion of unused areas and abundant space.

In other words mini park is a small outdoor space usually no more than ¼ of an acre. Most often they are located in an urban area surrounded by commercial building, houses, an small lots with few space where people can relax or enjoy.

Benefits of Mini Parks

1. It is important to note that parks are not intended to survive an entire city in the same way as neighborhoods are City Park.
2. It should be depend on the need of continuous community that is the nearby individual and families for whose use it was originally intended.
3. Support the overall ecology of the surrounding environment.
4. Protects and conserves local wildlife landscape and heritage.
5. Reduce pollution, traffic and consumption of resources such as oil.
6. Make community safer and more sociable.
7. Improve fitness and health.

OPEN SPACE

Open space is term of original concept of green city which reveals that there should be no thoughtless addition of building in the future. The three planning postulates sun, space and vendor should always remain our directive principle.

- A park is an open space area provided for recreational use. It is usually owned and maintained by a local government. Park commonly resemble open woodland type of landscape that human being find most relaxing. Grass is typically kept short to discourage insect pest and to allow for the enjoyment of the picnic and exporting activities.
- Trees are chosen for their beauty and to provide shade.
- The town level or city level parks should be ranged between 10-12m^2/ person. It includes parks play field, specific park, amusement park and multi-purpose open space (botanical garden, geographical park and traffic park, *etc.*).

Open Space Parks to the Institution Areas

As such areas the green open space parks are available either in the form of the private housing, schools, college and other institution in college campus. The plain green areas in city lay-out contribute as much as the overall city greenery and open space.

Major open space park

- Rose garden (Dr. Jakir Husain)-Chandigarh.
- Rock garden (spread over several acre)- Chandigarh.
- Rajendra park (400 acre)-New Delhi.

CITY PARK

Ornamental Trees Recommended for Town Roads

1. Shade Trees

Alstonia scholonis, Anthocephalus cadamba, Ficus infectoria, Azadirachta indica, Polyanthia longifolia, Putranjiva roxburghii and Tamarindus indica

2. Flowering Trees

Bauhinia purpurea, Bauhinia variegate, Cassia fistula, Cassia javanica, Cassia marginata, Jacaranda mimosoefolia, Delonix regia and Saraca indica

Landscaping Public Building

In the large city due to lack of space hardly any compound is left around these building for gardening but where ever space is available a lawn should be laid with few flower beds and foliage beds. Possibly same shrubs border may be added the entrance and exit roads may be lined with flowering trees, foliage plant and some hedge. As in educational institution garden, the aim of planting tree around building cut down the noise, control dust and provide shade.

Landscaping of Educational Institution

It is important that our administration and citizen are taking landscaping of campus seriously while planting any architectural designs for educational building, *e.g.*, Silver oak *Grevillia robusta, Polythea longifolia, Putranjiva roxburghii*, Seedling Mango (*Mangifera indica*) and Rain tree (*Samanea saman*). Plants should be suitable for planting in the front and front row and border planting, *e.g., Cassia fistula, Cassia nodosa, Delonix regia, Tecoma argentea and Bauhinia variegate, etc.* Where the electric wires limit the choice of avenue trees, small flowering tree such as *Callistemon lanceolatus, Bauhinia varigata* and *Tecoma argentea, etc.* can be planted. *Bougainvillea* and creeper over the wall of the building can change whole look of the building similarly a *Begonia venusta*.

Home Landscaping Basic Principal

1. Back garden
2. Contrasts
3. Balance or proportion
4. Open center
5. Repetition
6. Rhythm
7. Variety

LANDSCAPE ARCHITECTURE

Landscape architecture is professions which involve the design of the outdoor's. Its roots are in the gardening and agriculture, although this profession today has grown to overlap other fields such as planning architecture and arts, *etc.*

Elements of Landscaping Architecture

- ☆ Planting.
- ☆ Paving-walks, terraces, and playground.
- ☆ Shelters-overheads canopy and trellises.
- ☆ Screen and fences.
- ☆ Retaining walls.

- ☆ Elevation changes (steps and ramps).
- ☆ Bridges.
- ☆ Lighting.
- ☆ Furniture (Bench, chair, *etc.*).
- ☆ Information.

Principle of Design

1. Line
2. Form
3. Variety
4. Unity
5. Balance
6. Focal point
7. Rhythm
8. Colour
9. Texture
10. Scale
11. Proportion
12. Ratio

MUGHAL GARDEN

These gardens are of formal style orderly symmetrical planned and accommodated in rectangular or square plot. Mughal garden are a group of gardens built by Mughals in the Persian style of architecture. This Persian style was heavily influenced by the Persian gardens particularly the Charbhag garden structure. Significant use of rectilinear layout are made within the wall includes pools, fountain and canal inside the garden. The primitive plan design is simple in which plot is divided in to four equal plots by providing water channel in the centre. These channel are raised above ground and use for irrigation field. These water chamber are dividing in 4 river of life the plant are planted with trees and flowering plant. It's essential feature included running water (perhaps the most important element) and a cool to reflect the beauties of sky. Provided coolness and freshness to the garden the trees of various arts, some of them to provide shade mainly and other to produce fruit flower, colourful and sweet spelling and grass use under the tree in the garden.

Following points are kept in mind at the time of preparation of gardens:

1. **Site and design-** Mughal garden are generally rectangular or square in shape, it prefer the sloppy hills with a perennial plants or along the river bank.

2. **Wall and gates-** The creation of gardens by Mughals not only for pleaser and recreassion but also as forts and residence surrounded by high walls and with an imposing wooden gate at the entrance. The purpose of the high wall was security from enemies and shelter against hot wind.
3. **Terraces-** The basic plan is extended into terraces irrespective of topography of land the presence of 7-8-12 terraces in the garden symbolizes planets or paradise respectively on the contrary terraces are also made on flat ground by dividing into terraces.
4. **Running water-** The style for having running water by constructing canal and tanks was borrowed from the Persian. The water canals were made by blue tiles or marbles. The water canal used to have fountain to through up the water high in the air. The small lamp used to be illuminated to create beautiful reflexion in the evening.
5. **Baradari-** Baradari is made by stone with a pakka roof and used for sitting. It is a canopied building with twelve open doors, *i.e.,* three in each direction.
6. **Tomb or mask-** It was a common practice to have the garden built around a tomb. It is often shade that the mughal gardens were at their best when built around a monument.
7. **Trees and flowers-** The tree were selected with careful planning and thought. Each tree symbolized some thing like life, youth, death, *etc.* Fruit

Part of Mughal Garden at Rashtrapati Bhawan

trees were considered symbol of life and youth while Cypress represented death and eternity. The Mughal had bias for spring flowering trees and flowers. The favourite flowers were rose, jasmine, carnation, hollyhock, delphinium, *etc.*

Chapter 22

Avenue Planting

Landscaping of the national and state highway with trees is important aspect of beautifying our countryside. It is only one part of planting of trees on highway is necessary not only for the purpose of beautification but also for utility and necessity. The main purpose of road side trees is to provide shade during the summer. For wider roads doubles rows can be planted with other row's having shade trees and the inner row's of flowering tree.

The Mughals also planted trees on roadsides. They had highly paid staff, who neither trained in the subject nor had great love for planting with few exception. The road side trees on national highway should not have only to provide shade. Many of highway have been planted with trees such as tamarinds, mango mahua, guava and sesame, *etc.* The highway trees should never be planted in mix avenues but only one species should be planted for a long distance of the road. The trees should be planted 12m a part in row at least 5-6m away from the edge of row. If a road is as wide as 30m or more double rows of the tree should be planted row's being spaced 10-12m apart. If a road side trees is intended for timber, replacement should be done (5-6 years) at the cutting time. In rural areas the formal planting of avenues is done against the principle of landscaping.

Landscaping for Railway Station and Railway Line

One important aspect of bio-aesthetic planting is to landscaping public place on priority basic compare to private place. A well planned railway plot give a visitor or a passenger the first impression about the city. It will be a social obligation of our railway to keep the platform full of beautiful trees where waiting passengers can take rest in the shade of trees.

Railway Station

Besides flowering, foliage tree the railway authorities should also improve and beautify the platforms with palms and other attractive plant such as *Bougainvillea*. Even hanging basked can be displayed near the booking office or on the pillar on the resting shades in railway station or landscape avenues with flowering tree such as *Cassia fistula, Cassia nodosa, Delone regia, etc.,* The tourist will carry back pleasant memory from the city.

Railway Line

It will be a little hard task to landscape stretches of railway line compared the landscape of the platform specially in the drier track of the city. In India we have beautiful flowering trees such as *Beautia monosprma, Cassia fistula* and *Erythropsis calorata, etc.* The landscape of railway line has some problem such as watering. This can be partly overcome by planting drought resistant trees.

Ornamental Trees

Albizzia procera, Bauhinia variegate, Cassia fistula, Cassia javanica, Delonix regia, Jacaranda mymocifolia, Lager stoemia species, Polyanthia longifolia and *Saraca indica, etc.*

Economical Purpose

Anacardium occidental, Averrhoa carambola, Dalbargia sissoo, Mangifera indica, Shorea robusta, Tamarindus indica, Tectona grandis and *Terminalia arjuna, etc.*

Bus Terminal and Airport Landscaping

The bus terminal should be beautify by planting ornamental trees.

Along Bank-River and Canal

In India the rivers Ganga, Jamuna, Kaveri and Gomati, *etc.* are regarded as sacred and have many old Hindu temples and ghats. The bank along these places should be planted with flowering trees such as kadamb, ashok which are associated with Lord Krishna and Sita. Recently some good landscaping work has been done by the state horticulture department near the Gomati river at Lucknow by planting trees *Cassia nodasa, Cassia javanica* and Gulmohar, *Anthocephalus kadamba, Bauhinia varigata, Delonix regia* and *Saraca indica, etc.*

Landscaping for City Town and Countryside

Our city and town can be made more healthy and beautiful by resorting to bio-aesthetics planning. These can be achieved by planting roadside trees and establishing park, planted with green plants. Planting of trees on road side to provide cool shade which is much needed in the long hot summer months. Parks are places of relaxation, entertainment and fresh air in congested city. The broad road city should be planted with double row of trees. The outer road should be consisted of shade trees. Tall trees (eucalyptus and araucaria) with spreading habit (Banyan and *Ficus benjamina*) are not suitable for city and town roads.

Trees for Countryside

The trees planted for villagers should provide the fuel, timber, fruit and fodder. Babool is a common tree in the village. The bark of these tree is used for tanning leather and other purpose. Desi mango, jackfruit, seasam (*Dalbergia sissoo*), sal (*Shorea robusta*) for small tree langra, dashehri, chausa, palmirah palm, mahua and jamun, *etc.*, should be planted in villages. Foliage plants are valuable fodder for cattle (Babool, kachnar, neem and ber).

Chapter 23

Indoor Gardening

Indoor Gardening

The house plant or indoor plants have become a necessity of our homes today. Some Indian homes have to decorate a room by use of some plants for a period of time.

Selection of House Plant

The plant which are generally grown in the house are of two kinds. In the first category are flowering plant such as *African violets, Azaleas species, Geranium, etc.* which are spectacular in appearance by the virtue of their colorful flower. The other category provides permanent display with their graceful foliage and some time unusual architectural form, *e.g.,* bonsai. Though green leaf can also be very attractive especially if the shape is unusual or interesting, *e.g., Monster deliciosa* but leaf with some colour other than green are considered more attractive as in *Genera aurantiaca, Caladums species, Rex begonias, etc.* While other coloured forms of natural green leaf are available, *e.g., Pepperomia magnoliaefolia variegata, Ficus radicans variegata.* Beside these the ornamental foliage and flowering cactai, succulent, palms, ferns, bonsai and some bulbous plant can also be grown inside a house. A miniature water lily *Nymphaea pygmaea* can also be grown indoors in a bowl of 30cm diameter.

Cultivation

After a house plant procured it should be handled with care for the first one or two week when a plant is brought indoor directly from the favourable condition of a greenhouse. It shows sign of yellowing of leaf or even defoliation condition as a result of reduced light, less humidity and unfavourable temperature condition.

Light

It is an important factor in the cultivation of house plant while *Hedera helix* a climbing house plant thrives well under a relatively dark corner. Plant such as *Sensevieria trifasciata* requires a good amount of lights. A general recommendation is that 15-20 watts of florescent light is needed for each 30cm^2 plant area.

Temperature

The temperature required for house plant is much more variable than that of light because of divert climate condition of origin of different house plant. The ideal rang temperature should be around 15-21°C day and night temperature never below 20°C. The optimum temperature range in air for indoor plant is 21-30°C but most of house plant can hardly withstand higher temperature.

Humidity

Many of house plants, with few exceptions such as sensevieria, aspidistera desert cacti and few succulent in their national habitat, receive high amount of humidity. Relative humidity around 50-60 per cent is most ideal for indoor plant or house plant, such as alocasia, anthurium, ferns, orchids (80-100 per cent), *etc.* In miniature glass greenhouse the inside temperature is controlled thermostatically.

Watering and Fresh Air

An alocasia need more water than the cacti and some are polluted due to fumes high concentration of CO_2 which is detrimental for the most of the house plant.

Ideal Potting Media to Grow Indoor Plants

Sand : Silt : Leaf mould : FYM 1 : 1 : 1 : 1 (C:N ratio-1:200)

Modes of Indoor Gardening

The art of growing house plant or indoor plant inside a house for beautifying it is known as indoor gardening. The popular method of growing indoor plant are hanging basket, china basket, wood, cane, bowls and dishes, bottle, miniature gardening, window, gardening and vertical gardening, *etc.* There are some other methods of displaying house plant as living screen on a window by growing light indoor creepers such as *Hedera helix*, *Scindopus aureus* and golden pathos (money plant) or other plant.

Some House Plant

1. **Climbing-** *Ficus pumilla, Ficus radicans variegate* (Suitable for hanging basket), *Hedera halix, Philodendron elegans, Sundopsus aureus* (pothes), *Sundopsus aureus* (tricolor for creamy white variegate), *etc.*
2. **Trailing foliage plant-** *Chrorophytum comosus variegate, Zebrine pendula, Zebrine purpusii, etc.*
3. **Bush & upright foliage plant-** *Araucaria excels, Brassica actinophyla, Aglononema commutatum, Aralia elegantissima, Araucaria excels, Begonia spcies, Calthea nokoyana, Condeum species* (croton different colour), *Cyperus alternifolius, Dieffenbachia exotica, Ficus benjamina, Monstera diliciosa, Sensevieria trifasciata, etc.*
4. **Flowering house plant**
   a. **Climbing and traning type-** *Begonia glaucophylla, Begonia glabra, Trachelospermum jaminoides, etc.*
   b. **Bushy and upright type-** *Begonia maculate, Begonia manicata, Billbergia nutrians, Saintpaulia ionanth* (African violet), *etc.*
5. **Flowering pot plant-** *Azalea indica,* Tuberous rooted *Begonias species, Chrysanthemum species, Coleus species, Geranium species, etc.*
6. **Bulbus plant-** Daffodils, *Hyacinths species,* Tulip (Annual), Amaryllis (Perinials), *etc.*
7. **Ferns-** *Andiantum nephrolepis, pteris erotica, etc.*
8. **Palms-** *Chamacrops humillis, Cycas revolute, Phoenix roebelinii, etc.*
9. **Cactus and Succulent-** *Agave Americana variegate, Aloe species variegate, Cotylendon undulat Crassula arborescense, Opuntia microdasys, Sedum species, etc.*
10. **Bonsai-** *Cordia sebestena, Ficus diversifolia, Ficus nitida, Punica granatum, etc.*

Gardening in Tube or Urns

These are meant for display in the terrace, roof garden, backyard and on the doors steps. Wooden tubes are best suited for not sunny position, and rot resistant Wood such as teak, oak and popular, *etc.* are best suited for this purpose.

Culture 2-4 Kg cow dung or oil cake in 5 liters of water

Type of Plant

Annual- Marigold, chrysanthemum, sunflower, hollyhock, *etc.*

Herbaceous, perennial and bulbous plant- Cana, bird of paradise, datura, *Vinca rosea* and geranium, *etc.*

Shrubs- Lantana, camellia, azalea, *etc.*

Ferns- Birdnest (*Aspenium nidus*) *Cheilanthes argentea, etc.*

Others- Aralia, draceuna, araucaria, *etc.*

Topiary – *Duranta plumier, Baugainvilla* species, *Juniper* species, *Thuja species, etc.*

Gardening in Hanging Basket

Hanging baskets with trailing are cascading plants suited for indoor as well as outdoor. These can be hanged in the hall or drawing room besides a well little garden in the bathroom, above a florescent light, window gardening or window box, gardening in dishes, bowls and tray, jar, bottle garden, *etc.*

Section-II

Commercial Flower Production

Chapter 24

Rose

- Common name - Queen of Flower, Perfume of God
- Botanical name - *Rosa species*
- Family - Rosaceae
- Chromosome No. - 2n=14 (X=7)
- Ploidy level - Diploid
- Origin - South Asia
- National Registration Authority of Rose (NRAR) under International Registration Authority of Rose (IRAR) USA.
- Asia's largest Rose garden is Jakir Husain Rose Garden, Chandigarh, Haryana.
- Red colour of rose is due to anthocyanin pigment
- Orange red to scarlet colour of rose is due to pelargonidin
- Crimson to bluish red colour of rose is due to cynadin
- Blue to violet colour of rose is due to delphinine
- Yellow colour of rose is due to chorcones
- First book of rose "Rose in India" is written by Dr. B. P. Pal
- Scientific Rose Breeding in India is written by Dr. B.P. Pal
- Rose fruit is known as Hips
- First hybrid tea rose is "La- France" produced by Guillot (1867)

Classification of Rose

Rose Groups	***Parents***	***Varieties***	
		Indian	***Foreign***
Hybrid Tea	Hybrid perpetual × Tea rose	Purnima, Abhaya, Priyadarshani, Abhisarica, Pusa Sonia,	Superstar, First Prize, First Red, Paradise
Floribunda	Hybrid tea × Polyantha	Banjaran, Kumkum, Arunima, Mohini, Surya Kiran	Play Boy, Confety, Summer snow, Blue Berry Hill, Apricot Gem
Tea Rose	*Rosa chinesis* × *Rosa gigantium*		Anna Oliver, Lady Hill
Grandifloras	Floribundas × Hybrid tea rose		Queen Elizabeth
Damusk Rose	*Rosa gallica* × *Rosa phoenicia*		Celsiana

Introduction

Rose are symbol of beauty, fragrance and are used to convey the message of love. Without roses garden are not considered complete. The rose is one of oldest flower in cultivation and rank first position world wide as cut flower. There are two type of rose.

1. Stem less rose supplied by field grown rose for making garland or extraction of rose oil or rose water.

2. It is the stemmed rose which is mainly grown under protected condition for cut flower production used for making bouquets or interior decoration.

Genetics

Rose belongs to family rosaceae and genus *Rosa*. The genus contains about 120 species and is grouped under 4 subgenera viz; *Eurosa, Plotyrhodon, Hesperhodos,* and *Hulthemia*. The basic chromosome number of rose is 7 and some important species of Asian origin are diploid (2n = 14), whereas modern roses tetraploid (2n = 28).

Botanical Description

Rose bears terminal flower in general. In hybrid tea one or few terminal flower and in polyantha clusters of flowers in the inflorescence are found. Flowers are hermaphrodite with inferior ovary. In most species flower structure is based on 5 sepals and 5 petals, however the flowers are multi-petaled.

Origin and Distribution

Originally all roses are grown wild. About 120 known wild species of roses, all indigenous in the temperate region of northern hemisphere have been reported. The species growing wild in India are *Rosa brunonii* (Himalaya musk rose), *Rosa gigeantia, Rosa moschata,* (Musk rose) *Rosa microphylla, Rosa sericea,*(Ladakh rose).

Till the 19th century only 4 species *Rosa gallica* (Red rose), *Rosa canina* (Dog rose), *Rosa moschata* (Musk rose) and *Rosa Phoenicia* (Phoenician rose). Play a role in development of cultivars grown at that time. The French Rose (*Rosa gallica*) is an ancestor of moss roses. The Cabbage rose (*Rosa centifolia*) and Damask rose (*Rosa damascena*). The Persian musk one of the form of *Rosa moschata* is very common in India, it was introduce in England in 1599.The *Rosa gallica* known as French is native to Central and South East Europe.

Varieties

- ☆ **Hybrid Tea Variety-** Dr. Homi Bhabha, Diva Swapna, Apsara, Angar, Ganga, Lal Makhmal, Superstar and Happiness.
- ☆ **Floribunda Variety-** Ice-berg, Angara, Sadabahar, Pusa Barahmasi, Pusa Virangana, Sarada and Sukumari.
- ☆ **Polyantha Variety-** Preeti, Rashmi, Swati and Rekha.
- ☆ **Miniature Variety-** Delhi Scarlet, Dust Rose, Twinkle-Twinkle and Gypsy-Jewel.
- ☆ **Climber Variety-** Delhi Pink, Delhi White Pearl, Akash Pradeep, Pusa Komal, Pusa Satabadi, Pusa Manohar and Pusa Ranjan.

Soil and Climate

Sandy loam soil is ideal for the rose growing. Well drained and well decompost manure are used for rose cultivation which have pH range between 5.5 -7.

Rose requires good light throughout the year. These are sun loving plant capable of using high amount of light but at the same time fluctuation in light can result in leaf burn or petal damage. Temperature range is 15-28°C and high relative humidity around 75 per cent is ideal for good quality rose growing.

Propagation

1. **Cutting-** Matured current season shoot are selected for the cutting. The cut ends are dipped in root inducing hormones like IBA at 1000 PPM. Cuttings are gives roots in 2-4 week times.
2. **Budding-** The most commercially prefer types of rose plant material over world is the use of budded plant. Dormant eyes on a scion of chosen variety are budded either by "T" or "I" method of budding on a rootstock. The *Rosa multiflora* is most suitable rootstock for Southern part of India. In northern Indian condition or tropical region are mostly budded on *Rosa indica* and *Rosa manetti.*

Land Preparation

Summer ploughing should be done up to 30 cm deep so that the soil gets exposure to sun and air. If the soil is light sandy and stony the next 30 cm deep soil in the ground surface should be returned to the trench worked with digging fork and levelled.

Manures and Fertilizers

Basal dose of 15-20 tonnes/ha well be decomposed FYM should be applied before planting. Rose require 200 -400 kg Nitrogen/ha. This dose may be split into two doses once after pruning and 2nd dose after about 30 days. The requirement of P and K can be apply 150kg/ha of each after pruning.

Planting

Favorable time of planting is June to October where winter is severe. Planting can be done in autumn or spring season. Planting distance is 75 × 75 cm and 60 × 60 cm with flood irrigation facility or 30 × 30 cm, 45 × 30 cm and 45 × 45 cm in single, double and triple row system in drip irrigation facility. Planting should be done in well prepared bed and during planting the soil should neither be too wet nor dry. The earth ball is scraped lightly and gradually lowered in the hole and allowed to rest on its bottom. The gap in the hole is then filled up the pressed properly to anchor the plant firmly.

Irrigation

The frequency of irrigation depends on many factors such as growth, soil texture, climate and field condition, *etc.* Maintain adequate soil moisture at all stages of its growth and at flowering time. Irrigation is good just after pruning and the excess of irrigation during flowering period may be harmful. Deep irrigation should be avoided.

Cultural Operation

Pruning

For cut flower production pruning is done during first week of October in north Indian condition whereas in last week of June and last week of November in Bengaluru. However in case of essential oil bearing varieties of *Rosa damascena* pruning is done from last week of December to the beginning of January. The height of pruning varies from 30 – 45 cm from ground level.

Mulching

Mulching conserve soil moisture, supplies humus and suppress weeds. About 15 – 20 cm of mulching with paddy straw, dried leaves of sugarcane and tree leaves in rose beds reduces weeds.

Harvesting

For cut flowers, stems are harvested in the morning at tight bud stage or two petals begin to unfold. Stem should be cut leaving at least 2-5 leaflets on stems and should immediately be placed in bucket containing water. For loose flower which is generally used for making garlands, flowers are plucked at half open or fully open stage depending on situation and market distance. For making rose water or oil newly opened flowers are harvested in the morning hour before sun rise and distilled immediately.

Yield

About 10-20 flower stems per plant can be obtained in hybrid Tea roses grown under open condition for cut flower purpose. On an average *Rosa damscena* produce 20-50 quintal/ha. 100-200 flower for loose flower. Flower weighs 2-5 gm. For protected condition 24-26°C day temperature, 15-17°C night temperature, humidity 75 per cent on an average yield 180-300 flower/plant/year.

Post-Harvesting Management

After harvesting the flower should be shifted to cold room to remove field heat from the flowers. Slowing down of respiration lowers water loss and arrests the excessive opening of bud.

Floral Preservative

Vase life of roses can be increased by using floral preservative Aluminum sulphate (300 ppm) in vase solution. Pulsing of roses with sucrose (3 per cent) +Aluminum sulphate (300 ppm) increase of vase life.

Grading and Packaging

Long stemmed varieties are graded from 40cm onward with the differences of 10 cm while short stemmed varieties are graded from 40-60cm with the differences of 5cm. The cut stems should be made in to bundles of wrapped with two play soft corrugated paper to secure the buds in position.

Insect Pests

1. **Mite** – This is a polyphagous pests which are mostly found in lower side of the leaves covered with fine silky webs. Due to their feeding, white specks appear on the leaves and specks coalesce and appear as white patches. Abamectin (0.5 ml/lit), Difenthiuron (0.5 ml/lit) should be sprayed to control it.

2. **Thrips** (*Thrips fuscipennis*)– Leaves may also show silver flecking and brown thrips are usually present on plants. Control by Malathian (1ml/lit) sprays should be made soon after the early infestation with a repeat application after 2 -3 weeks if damage continues.
3. **Jassid** (*Odontotermes obesus, Microtames obesi*)– Jassids sucking insects cause yellowing or whitening of the attacked surface. It can be controlled by metasystox or democron (0.1 per cent) or soil application of thimet (0.1 per cent).

4. **Aphid** (*Macrosiphum rosae*)– Large, dark green or pink brown aphids feed on buds, shoots and leaves. It can be controlled by Dimethoate-B (0.15 per cent).

Fungal Diseases

1. **Die back** (*Botrydiplodia theobromae, Colletrichum gleosprroids, Fussarium* spp.)– The disease causes the death of the plants from top to downward. The diseases start from the pruned surface of the twigs. Soil drenching with 2gm/lit Bavistin or Benomyle are effective to control the disease. Spraying with Mancozeb or Copper oxychloride 2gm/lit immediately after pruning and then twice and 10 days' interval is effective.

   Die back tolerant varieties are in hybrid tea Rose Arjun, Bheem Dr B.P. Pal, First Prize and Abhisarica, Priyadarshani and in Floribunda Arunima, Delhi Princes, Jantar-Mantar, Lahar, Madhura, Sada Bahar, Suchitra, Usha.
2. **Black spot** (*Diplocarpon rosae*)– The disease is also called as leaf blotch, leaf spot, blotch and star sooty mould. Root stock like IIHR thornless and IARI thornless are tolerant to disease. This disease is controlled by Spray of Ferbam (0.1 per cent) at fortnightly interval, Benlet or Bayleton (0.1 per cent) apply just the appearance of spot.
3. **Powdery mildew** (*Sphaerothica pannosa variety Rosa*)– It is major disease of rose spread over world. The younger leaf curl, exposing the lower

surface and such leaf are light are lightly to be purple to the normal leaf. Mostly occurs during November-March having its maximum intensity till June- February. *Rosa multiflora* tolerant to powdery mildew. Disease resistant varities are Gladiator, Raktgandha, First-Prize, Eiffel Tower and Oklahoma.

4. **Botrytis blight** (*Botrytis cinerea)*– The petals are affected, buds develop browning spot which shown, entire surface and causes rotting. This disease is controlled by spray of Bavistin (0.2 per cent), Benomyl (0.2 per cent) and Rovral (0.2 per cent).

Physiological Disorders

1. **Blind shoots–** Blind shoot may be defined as a stem that fail to develop a bud. Low photosynthetic accumulation rate results in up to 57 per cent of shoot developing into blind shoot. In a study it was found that temperature lower than 15°C increases number of blind shoot. These blind shoot pinched back hard by one or two nodes in November resulted in flowering stem. Lighting with sodium lamps is also found to decrease the number of blind shoots significantly. Foliar spray of ascorbic acid @1000PPM also reduces blind shoot.
2. **Bent neck–** It is characterized by the bending of stems of cut rose flower after harvesting and is an important factor in determining post-harvest quality. In the bent neck result to soft growth, premature bud harvest and excessive water loss during handling, exposure to high temperature, low relative humidity, ethylene and high microbial growth bent neck increase. Use of 200 PPM copper nitrate along with 10 per cent sucrose in floral preservative at pH 6 is found to be very effective in controlling bent neck.
3. **Bull head–** The rose flower malformation commonly known as bull head. The term bullhead is described as the flower which fails to open properly has short petal or an excessive number of petals. Bull head is commonly found in roses observed mainly in low temperature during night. It occurs due to an abnormal production of cytokinins and gibberellins hormones responsible for cell elongation and stem elongation.
4. **Balling–** The inability of a bud to open into a bloom due to excess moisture which causes the petals to stick together. It occurs in areas with cool and damp night.

Chapter 25

Dahlia

- ✰ Botanical name - *Dahlia variabilis*
- ✰ Tree Dahlia - *Dahlia imperialis*
- ✰ Cactus Dahlia - *Dahlia jaurezi*
- ✰ Family - Asteraceae
- ✰ Chromosome No. - 2n= 16, 32, 64 (X=8)
- ✰ Origin - Mexico
- ✰ Dahlia is deciduas, tuberous rooted, hardy and perennial flowers, facultative short day to day natural plant.
- ✰ Dahlia first introduced in India in 1957 by Agri. Horticulture Society of India, Kolkata.
- ✰ Dahlia is commercially propagated by terminal stem cutting and other by seed, tuberous root.
- ✰ Planting time of dahlia in north India is September- December and in south India May- June.
- ✰ Largest producer of tuberous dahlia is Netherland.
- ✰ Dahlia is also known as King of Flower

Introduction

In modern garden dahlias are extensively used for exhibition, garden display and decoration. It grows well in pot also. For border purpose, dwarf growing type are preferred while in pot usually large flowering dahlia are grown. The long stemmed flowers of various form and colours are used in flower arrangement. The tubers of dahlias contain appreciable amount of insulin and fructose and benzoic acid which are off medicinal and nutritional value.

Classification

According to National Dahlia Society of England:

1. **Single flowering -** One single of florets, central group of disc floret suitable for bedding purpose. Height are 40-60cm, *e.g.,* Bambin Little Dorite, yellow hammer.
2. **Anemone flowers-** Fully double one or more ring of florets, good for flower arrangement. Height are 60- 90 cm, *e.g.,* Comet, Gunia, Scarlet comet.
3. **Collerette-** One outer ring of flat florets an inner ring of collar florets and central groups of disc florets. Hight 75-120 cm good for flower arrangement, *e.g.,* Choh, Sincerity and Thais.
4. **Water lily-** Fully double flattened shape florets are flat with slightly curved margin popular for cut flower up to 120cm, *e.g.,* Porcelain, Jescot Lynn and Christopher Taylor.

5. **Decorative-** Fully double height about 150 cm. It is of 5 types which are given below.
    a) **Giant decorative –** The most popular normal tendency is to grow them as large as possible, *e.g.*, Bonaventure, Hamari Girl, White Alvas, and Edge of Gold.
    b) **Large decorative-** Very popular in India, *e.g.*, Black Out, Polyand, Silver City.
    c) **Medium decorative-** Having bicolour flower very good for garden display and flower arrangement, *e.g.*, Alloway Cottage, Daleko National, Evelyn Foster.
    d) **Small decorative –**Numerous cultivar including bicolour are good for garden display and flower arrangment, *e.g.*, Corton Linda, Disneyland, Lady Linda.
    e) **Miniature decorative-** Most popular for flower arrangement and garden display and good uses for vase life, *e.g.*, Christine Elizabeth, Hammet, Eastwin.
6. **Ball group** – Fully double, ball shaped (after flat) florets are blunt or round ended height about 120 cm. Ball are of 2 types.
    a) Small ball- Alltamy Cherry, Risca Miner.
    b) Miniature ball- Nettie, Rothsay Superb.
7. **Pompon decorative-** Fully double globe shape, involutes florets are blunt or round ended. Height 80-120 cm popular for the flower arrangement and longest uses of vase life among the entire dahlia, *e.g.*, Diura Gregory, Hallmark and Norien.
8. **Cactus-** Fully double involute floret are narrow and pointed. Height is 150cm. Cactus are of 5 types.
    a) **Giant cactus-** Poler Sight, Donna Huston.
    b) **Large cactus –**Eastwood Show, Golden- Crown (exhibition purpose).
    c) **Medium cactus –**Sunset (arrangement).
    d) **Small cactus-** Alvas Doris, Doris Day (vase life).
    e) **Miniature cactus-**Rokasley Mini, Frant Soetan (for cut flower).
9. **Semi-cactus –** Fully double pointed floret are involutes for half their length or less, Height about 150 cm. It is 5 types
    a) **Giant semi cactus** –Gate Way, Daleko Jupiter (exhibition purpose).
    b) **Large semi cactus-** Hamari Princess, Reegimald Keene and De Sarsate.
    c) **Medium semi cactus** – Symbol and Amelis Neerd.
    d) **Small semi cactus** –Match, Kimono, Shandy.
    e) **Miniature semi cactus** – Snip, Marry Jo, Snow Queen.

10. **Peony flowering** – Fully double, floret are round ended, height 100 cm, *e.g.,* Bishop of Liandaff, Fascination.
11. **Dahlinova** – Double flowering variety in every available color. Height is 20-30 cm, the tubes denser and smaller than those in the normal assortment, *e.g.,* Dahlinova Arizona, Dahlinova Virginia.
12. **Gallery series** – This series contains cactus and decorative variety which become no taller than 35 cm, *e.g.,* Gallery Rembrandt, G. Art Deco.
13. **Impression selection**- These are Collarette dahlias and primarily suitable for bedding and use for balkani, height varies farm 30-50 cm, *e.g.,* Festivo, Impression Fortuna and Impression Fuego.

Varieties

1. **Giant Decorative**- Blood Orange (Orange and Red), Dick West Fall (Pink and yellow), Alavas Supreme (yellow), Barbara Marshal (Glowing red), Bonaventure (yellow bronze blend colour), Croydon Ace (Deep yellow).
2. **Large Decorative**- Alden Galaxy (Red), Blank Out Sport (variegation, rosy red with blackish red strips over Patel) Islander (Deep pink), Queen Elizabeth (Lemon Yellow).
3. **Medium Decorative**- Avoca Pawnia (yellow), Geerling Moon light (yellow), Trengrove Jubilee (yellow), April Down (pink and white bland), Bhikuss Mother (Bicolour orange to tan with white tip).
4. **Small Decorative**- Camano Choice (Yellow and pink blend), Corton Linda (white), Disney Land (Bland of yellow, red, bronze colour), Lady Linda (yellow type with lavender), Nina Chester (white).
5. **Miniature Decorative**- Rye Croft Gem (lavender), Rye Croft Pride (purple pink), Rural Dirk (yellow), East Win (purplish red).
6. **Semi cactus**- Luther (creamy white), Carol Chaning (bronze), Daleko Jupiter (red and white) Davenport Sunlight (yellow), Date Way (white), Mandarin Napoleon (orange).
7. **Cactus dahlia**- Alva's Dories (bright red) Banker (flame red), Camano Classic (orange and bronze blend), Doris Day (scarlet colour), East Wood (snow white).
8. **Water lily dahlia**- White Cameo (white), Christopher Taylor (pink).
9. **Ball dahlia**- Green Way Joy (red purple), Camano Candy (pink), Risca Miner (purple), Senior Boll (lavender blend) and Snow Fall (white).
10. **Pompon dahlia**- Diano Gregory (lavender), Hallmark (Pink), Iris (purple), Little Snowdrop (white), Noreen (pink and purple), Moor Place (Purple).
11. **Mutant variety**- Prabodh, Swamiji, Bhikku, Manju Shree, Zail Singh.

Soil and Climate

Dahlias are grown under the well in loamy soil as well as drained sandy soil with some organic matter. It is cultivated at the good pH range between 6-7.

Best quality plants were obtained when grown in 25°C day and 16°C night temperature but flowering delay at 24°C day and 12°C night temperature. Although flowers are associated with 29-30°C day and 17-20°C night temperature plant quality was adversely affected.

Propagation

a) **Division-** Multiplication by division of tuberous root can be done by separating the tubers each with a piece of stem. Division of tuberous roots is easy and most satisfying way for amateurs.

b) **Cutting-** Commercially it is propagated by stem cutting, for this purpose the selected tuberous roots are held over from the previous season. Terminal stem cutting of 7-8 cm length with 2-4 leaves are prepared and cut part treated with IBA 1000 ppm and planted in washed coarse sand for rotting. The cutting normally roots within 10-15 days.

c) **Seed**- September-October is proper time of the sowing; seed should be covered with 2-2.5cm of soil after sowing. The favorable temperature is 18-28°C for seeds germination. Seedlings grow quit rapidly and become ready for the transplanting within 20-25 days.

Land Preparation

Dahlias can be grown in pots as well as in ground. It can do well in both conditions. Any type of soil which is well drained and fertile is suitable for the cultivation of dahlia. Less fertile soil is more preferable for tuber production. During land preparation, after digging the ground, a thick layer of pulverized cow dung manure about 10 cm thick should be spread followed by the addition of 60 g SSP 30g sulphate of potash and 150g bone meal/m^2.

Manures and Fertilizers

Balanced feeding is essential for successful cultivation of dahlia. For increased vegetative growth and improved flowering 100 kg N, 125 kg P_2O_5 and 100 kg K_2O/ha is optimum. Full dose of phosphorus and potash and half dose of nitrogen should be applied during final land preparation before planting and remaining half dose of nitrogen after 40 days of planting. The foliar application of ammonium dihydrogen and potassium nitrate 0.3 per cent solution at weekly interval is beneficial for production of richly coloured large size boom.

Planting

Dahlia behaves as summer flower in hill while in the planes region winter flowering. Sprouted tubers are mostly use for planting in the hills while fruited stem cutting are used in the field. In hill dahlia are planted during April-May. The planting of cutting is done September- November (Kolkata) September-December (all). The planting distance will vary according to the grown habit that is 75cm in case of tall, large type 60cm for medium and 30-45 cm in case of dwarf varieties.

Irrigation

Dahlias requirement of water is more than other flowers. Applying water judiciously during a severe dry spell is found beneficial. Plants grown in pot should be watered daily and the plants on the ground at 3 to 4 days interval.

Cultural Practices

Weed Control

Manual weeding, mulching with grass chipping, straw, sawdust, black plastic are very beneficial. Sethoxydim at 0.28 to 0.47 kg a.i./ha is applied on the standing crop should be incorporated 5-6cm deep for annual weeds and 18 cm deep to suppress the growth of certain perennial weeds.

Training

When dahlia plant grows from sprouted tuber, only one strongest shoot should be retained and other are removed. To force the branches below the ground level a pinching is done as soon as 2-3 pair of leaf have formed.

Staking

Most dahlia growers provide some support to plants, except for low growing bedding varieties as their stems are very brittle and liable to break and hence, these should be supported by sticks, bamboo canes or similar strong stakes may be used for providing support to the soft and heavy stem.

Pruning and Disbudding

Pruning and disbudding to keep the center of bushed open. Large decorative varieties should have 4-5, while small flowering type may carry 8-10 branches.

Harvesting

Flowers should be cut early in the morning or in evening at a stage when they are fully open. After harvest stem must be emerged in water to 50°C for 30 second to prevent blocking of water canal.

Yield

Pompon dahlias are used for the cut flower. Yields are approximately 8-10 flowers per plant and in case of loose flower 15-30 flowers per plant.

Insect Pests

1. **Aphid-** Three species of aphids infect dahlias viz. Bean, green peach and leaf curl plum. Spray malathion (0.2 per cent) as soon as the aphids appear.
2. **Leaf hopper** (*Empoascus fabae*)- These pests causes discolouration of leaf, which appears first along one margin. The affected area is at first pale yellowish, later it becomes brown and brittle. Spray with Metasystox and Malathian (0.2 per cent) is very effective.
3. **Thrips** (*Heliothrips haemorrhoidalis, and Thrips tabaci*)- These infest the flowers. They rasp surface and feed on the exuding juice. Under surface of petal turn whitish and wither. Spray of Metasystox and Malathion (0.2 per cent) is effective.

Namatode

1. **Root knot nematode-** Two root knot nematodes (*Meloidogyne incognita* and *Meloidogyne hapla*) occur on the roots of dahlias when plants are grown in infested soil.

Fungal Diseases

1. **Powdery mildew** (*Eryspihe cichoracearum*)- The entire foliage is covered with a white powdery growth. It can be control by spray of Carbendazim, Quntozene, Benomyl @ 0.2 per cent.
2. **Stem root** (*Schlerozene rolfsii*)- The stem rot fungus attacks the main stem and branches. Affected plants should be uprooted and burnt. The tubers should be treated with Benomyl (0.1 per cent) for 30 minutes.
3. **Wilt** (*Fusarium* species)- Plant wilt and die because of blockage or destruction of conducting vessels. It can be controlled by removal of infected plant.

Bacterial Disease

1. **Crown gall** (*Agrobacterium tumifaciens*)- Infected plants becomes stunted and shoots becomes spindly. It can be controlled by dipping the roots in streptomycin solution.

Viral Diseases

1. **Mosaic-** Mosaic is very serious disease in dahlia it is caused by *Marmor dahliae* and it is transmitted by aphid. Transmitting agent can be controlled by spray of rogar 2ml/lit water in weekly interval.
2. **Spotted wilt-** This melody is characterized by motling of leaves stunting and ring spot symptoms. It can be controlled by application of Malathion (0.1 per cent).

Chapter 26

Tuberose

- ☆ Common name – Rajanigandha
- ☆ Botanical name – *Polyanthes tuberosa, Polyanthes durangensis, Polyanthes geminiflora, Polyanthes graminifolia, Polyanthes longiflora, Polyanthes montana.*
- ☆ Family – Amaryllidacae
- ☆ Chromosome No.- 2n=30 (X=15)
- ☆ Origin – Mexico and Asia Minor
- ☆ Propagated by - Bulb

Introduction

Tuberose is the one of most important tropical and subtropical ornamental bulbous flowering plant, cultivated for production of long lasting flower spike. It is mainly known as Rajanigandha or Nishigandha. Tuberose is an important commercial cut flower as well as loose flower crop due to pleasant fragrance, longer vase life of spike, higher returns and wide adaptability to varied climate and soil. Tuberose blooms throughout the year and it's cluster spikes rich in fragrance, florets are star shaped waxy and loosely arranged on spike that can reach up to 30-45 cm in land. Single varieties are more fragrant than double type of flower which is used in high grade of perfume. Flowers of the single type (Single row of perienth, single row of repels) are commonly used for extraction of essential oil, loose flower, flowers making garland. While double flower varieties (more than three rows of perienth or petals) are used as cut flower garden display and material decoration. Cultivation of tuberose is gaining momentum day by day in our country.

Botanical Description

Tuberose is a half bulbous plant, it is perpetuated through its bulblets. Bulbous are developed from scale and leaf bases. Numerous lenceolate leaves are green, narrow, linear, long and arise a rosette. The flowers have funnel shaped perianth, waxy white about 25 mm long, single and double borne in spike.

Origin and Distribution

Tuberose is native of Mexico and spread to different part of world during 16th century. The generic name *Polyanthes* is derived from Greek word *Polis* which means white and *Anthos* means flower.

Type of Flowers

On the basis of number of row of petals tuberose varieties are classified in to 4 types.

1. **Single flower tuberose-** Varieties having flower with one row of corolla segments. Flowers are extensively used as loose and extraction of oil, *e.g.*, Single Maxima, Kalyani Single, Shringar (It is cross between single into double) released by IIHR which are used for loose flower yield 15000kg/ha/year. It is tolerant to root knot nematode. For example Prajwal (it is

cross between Shringar in to Mexican Single) released from IIHR and used for loose and cut flower, Rajat Rekha (NBRI) (It has silvery white stick along the middle of leaf suitable for beautification), Hyderabad Single, Kalkatta Single, Phule Rajani and Pune Single.

2. **Semidouble flower tuberose** - These flowers have 2-3 rows of corolla segments. Spikes are straight, flowers are white and generally cultivated for cut flower purpose.
3. **Double flower tuberose** - The floret of this flower cultivar has more than 3 rows of petals. Spikes are used as cut flower as well as loose flower and for extraction of essential oil. Pearl Double, Kalyani Double, Swarna Rekha (Its bears double type of flower and have golden yellow sticks along the margin of leaf it has high aesthetic importance), Suvasini (Crossed between Mexican Single in to Pearl Double) give 25 per cent more flower yield over normal double cultivar. Its spikes are best cut flower Vaibhav (It is cross between Mexican Single into IIHR-2). The flower buds are greenish in colour and spike yield as higher to Suvasini variety best suitable for cut flower and pot purpose, Arka Nirantra (It is higher spike yield early white single flower) are some important variety of double flower tuberose.
4. **Variegated flower tuberose** - In this margin are variegated and highly suitable for land scalping. Variety silver whither or golden yellow strikes are visual on the leaf, *e.g.,* Rajat Rekha and Swarna Rekha (by radiation).

Soil and Climate

Tuberose can be grown in wide variety of soil including saline and alkaline soil. Tuberose performs well in well drained fertile, friable soil rich in organic matter, loamy and sandy loam soil with good aeration and drainage. The ideal soil pH is 6.5 - 7.5.

Tuberose perform well in mild climate without stream low or high temperature. It grows profusely from April-November and optimum temperatures are 20-30°C. The low temperature below 10°C is adversaly affects the growth and flowering of the plant.

Propagation

1. **By Bulb** - Tuberose is commercial propagated by bulb. Bulb should be spindle shaped, free from diseases and having and average diameter of 1.5-3.0 cm and 1.25-1.50 lakhs bulbs (8-9 tan)/ha. Bulb should be treated with 4 per cent thiouria.
2. **By seed** - Seed propagation is mainly followed by breeder to evolve new variety. Under favourable climatic condition seed setting is observed only on single flower cultivar. Germination is mainly affected by temperature and moisture. Soil temperature of 26.6°C is optimum. Seed should be sown 10cm apart in row and 2cm deep in light soil. Seeds start germinating within 10-15 days after sowing.

Spacing

Bulb should be planted at 30cm row to row and 20cm plant to plant with 7cm depth. It will depend upon the diameter of bulb.

Land Preparation

Summer ploughing should be done up to 30 cm deep so that the soil gets exposure to sun and air. If the soil is light sandy stony the next 30 cm deep soil in the ground surface should be returned to the trench.

Manures and Fertilizers

Well managed tuberose crop require 200 kg N, 150 kg P_2O_5 and 200 kg K_2O per ha. The half dose of nitrogen full dose of phosphorus and potash should be applied at final preparation of land. The rest dose of nitrogen should be applied in 3 split doses, *i.e.,* 30, 60 and 90 days after planting.

Planting Season

Well developed spindle shaped bulb with diameter 1.5 cm and above farming at the outer periphery of the clump are considered ideal for planting. In north India tuberose are generally planted in February-April (plane region) and in April-May (hills region). In south India bulb is planted during July-August.

Irrigation

The irrigation in tuberose largely depends upon soil type, stage of growth and climatic condition. In general light soil require more irrigation whereas in heavy soil require less irrigation. In summer month the crop should be irrigated at 5-7 days interval and during winter month at 10-12 days interval.

Cultural Operation

Weed Control

The pre-emergence application of Pendamethaline 1.25kg/ha, reduces the weed population. The pre-plant application of Antrazin at 3kg a.i./ha is effective in weed control. Manual weeding is beneficial but it is time taking and costly process.

Harvesting

1. **Flower-** Flowering of tuberose start 2-4 month after planting and flowering time is July onward. August-September is the peak period of flower and flower all the year around. The tuberose is harvested by cutting the spike when 1-2 pair of flower opens in single type and 2-4 pairs of double type varieties. About 4-6 cm basal portion has to be left to allow the growth of the bulb.
2. **Bulb-** Bulb needs maximum 7-8 month for proper maturation. At this stage the old leaf become dry and bulbs are almost dormant. The leaf of

clumped or clipped and bulb is stored at 30°C for 6 weeks. Irrigation of field few days prior to uprooting and soil is allowed to dry before digging out the bulbs.

Yield

Normally 3-5 lakhs spikes/ha cut flower or 10-15 tonnes/ha loose flower can be obtained. Yield of bulb varies from 20-30 ton/ha end of 3rd year.

Post-Harvest Management

1. **Floral preservative-** Vase life of tuberose can be increased by using floral preservative Aluminum sulphate (300 ppm) in vase solution includes vase life.
2. **Pulsing-** Pulsing of tuberose with sucrose (3 per cent) +Aluminum sulphate (300 ppm) increase of vase life.
3. **Grading and packaging-** Long stemmed varieties are graded from 40cm onward with the differences of 10 cm while short stemmed varieties are graded from 40-60cm with the differences of 5cm. The cut stems should be made in to bundles and wrapped two soft corrugated paper to secure the buds in position.
4. **Lifting and curing of bulb-** At the time of complete maturity, leaf become yellow and dry. At this stage irrigation is and after 2 weeks clumps of bulb are dug out from the ground. Clean and graded bulbs are stored on shelves for curing. The positions of, these bulbs should be changed after 2 week to prevent fungal attack and rotting.

Insect Pests

1. **Grasshopper-** The infestation of grass hoper is more during rainy season. Insect feed the young leaves and flower bud. It can be controlled by application of Malathian or Rogar 1ml/litter water at 15 days interval.
2. **Aphids-** This is the sucking type pest. Both nymph and adult suck the sap of flower buds and growing point. It can be controlled by spraying of Rogar at 2ml/litter of water.
3. **Thrips-** Thrips suck the sap from tender shoots, flowers and floral stock. Thrips can be controlled by application of Monocrotophas 1.25ml/litter water.
4. **Weevil-** Weevils are the soilborne insect. It is more active in darkness, damage the shoot and feed the leaves. It can be controlled by application of folidol dust at the time soil preparation before planting.

Nematode

Nematodes like Root knot nematode *Melydogyne incogneta* Rainy farm nematode (*Rotylenchulus renifarmis*), Foliage nematode (*Aphelelechoides besseyi*) and also greasy strike caused by *Aphelelechoides besseyi* found to wipeout of tuberose flower production. In severe infestation emergence of spike is suppressed. Application of furadon 2g/plant control the nematode infestation. The intercropping of marigold is also beneficial for controlling the nematode.

Fungal Diseases

1. **Stem root** (*Schlerotium rolfsii*)- The main symptom of stem rot is appearance of prominent spot of loose green colour due to growth of mycelia mosses on leaf. Leaves become yellow, droop and ultimately dry up. It can be controlled by application of Bavistin 0.2 per cent, three spray at 20 days interval and soil application of brassical @30kg/ha.
2. **Leaf blight** (*Botrytis elliptica*)- This disease can be controlled by application the plant with ammonical copper 0.2 per cent, carbendazim 2gm/litter at 15 days interval.
3. **Alternaria leaf spot** (*Alternaria polyanthi*)- It is the prevalent in rainy season. Leaves and peduncle become necrotic and dry up. It can be controlled by Mancozeb 0.2 per cent or Bourdex mixture 0.4 per cent at 10 day interval.

Bacterial Disease

1. **Flower bud rot** - This disease is mainly appeared in young flower buds and result in dry rooting of buds with brown scorch necrotic discolouration of peduncles. This disease can be controlled by spraying of Streptomycin 0.01 per cent.

Chapter 27

Chrysanthemum

- ✰ Common name: Glory of east, Guldawadi, Autumn queen, Queen of east, National flower of Japan
- ✰ Botanical name - *Chrysanthemum species/Dendranthema grandiflora*
- ✰ Family - Compositae/Asterace
- ✰ Chromosome No. - 2n=36 (X=9)
- ✰ Origin - China, Europe and Asia
- ✰ Small type, Double Korean type, mostly grown in open field condition.
- ✰ Commercially propagated by suckers and terminal cutting (June).
- ✰ *Chrysanthemum cinerariifolium* inflorescences are of considerable importance in the manufacture of pyrethrin insecticides.

Introduction

The *Chrysanthemum* is one of the most beautiful and perhaps oldest flowering plant. In Dutch cut flower action, chrysanthemum rank 2nd after rose. It is the one of most important traditional flower of India mainly used as a potted plant, loose flower, cut flower and as border plant in the garden.

Genetics

Chrysanthemum is the most photo-genetically advanced dicotyledonous family. Florist chrysanthemum - (*Chrysanthemum morifolium*), cut flower type (*Chrysanthemum maximum*) and for pot plant (*Chrysanthemum frutescens*). Garland chrysanthemum winter season annuals (*Chrysanthemum coronarium*) are grown in temperate region.

Botanical Description

The species of the chrysanthemum are annual and perennial herbs, sometimes partly woody. Leaves are alternate and inflorescence heads are many flowered.

Classification

Classification based on kind and arrangement of florets into two categories by National Chrysanthemum Society England–large flowered and small flowered. Large flowered are classified into 13 classes and small flowered are classified into 10 classes.

A. Large Flower

1. **Regular Incurve-** Ray florets are narrow to broad florets from a perfect ball. Average bloom size varies from 10-15 cm and disc not visible, *e.g.*, Snowball, Autumn King, Chandrama, Sonar Bangla, Super Giant and Adventure.

2. **Irregular Incurve-** Ray florets are broad and smooth. The bloom size is very large and varies from 15-20 cm. Disc florets are entirely covered by the upper florets, *e.g.,* J. S. Salisbury, Mountaineer.
3. **Skirted Incurve-** Bloom size varies from 15-20 cm. The lower florets mostly basal florets are bending downward in an irregular fashion to give a skirted shape, *e.g.,* Dream Castle.
4. **Incurving-** Ray florets are incurved upward in indefinite manner. Loose fitting arrangements of florets are present, *e.g.,* Classic Beauty, Dr. S. Mukharji, Leading Lady and Pink Cloud.
5. **Reflex-** Bloom size is 15-20 cm and ray florets are narrow to broad and bend to backward and downward, *e.g.,* Imperial, Rose Day, Star of India, Kasturba Gandhi. This class are found 3 types.
   a) **Regular reflex-** Ray florets are bending back and downward in a regular arrangement.
   b) **Irregular reflex-** Ray florets are bending downward in a twisting and irregular way.
   c) **Reflexing reflex-** Bloom appears flat in shape, *e.g.,* Imperial, Rose Day, Star of India, Kasturba Gandhi.
6. **Intermediate-** The blooms falling are intermediate between incurved and reflex and inner florets are incurved. The blooms are about 15 cm or more are globular shape, *e.g.,* T-1 and Sun Flight.
7. **Ball-** The ray florets are straight and densely packed radiation uniformly in all direction to give bloom a ball, *e.g.,* Pride of Madford and Nigeria.
8. **Quilled-** Ray florets are tubular and elongated with tips open or closed, *e.g.,* Green Sensation and Pradhans Prides.
9. **Spider-** The florets are large tubular and elongated and tips are coiled may be open or closed, *e.g.,* Gitanjali, Achievement, Innocence, Miss Universe and Diamond Jubilee.
10. **Spoon-** Bloom having visible disc ray florets are tubular with spatula like open tips *e.g.,* Pink Casket, Carnation, and Pusphahans.
11. **Anemone-** The ray florets are ligulate or quilled. Disc is hemispherical are raised, *e.g.,* Cloud Bank, Red Admiral.
12. **Single-** Ray florets are long, elongated and strap like, *e.g.,* Joan Helen and Surja.
13. **Semi double-** Ray florets are long, elongated and strap like. Number of whorl or florets are more than five, *e.g.,* Ronald, Crimson Tide.

B. Small Flower

1. **Anemone-** Disc florets are well developed and prominent while florets may be twisted and quilled, *e.g.,* Venus, Pink Cushion, Ace, Mercury.

2. **Button**- Button are small and compact 2-3 cm diameter, florets are short rayonate like and hemispherical radiating in all direction, *e.g.*, Lilliput, Kingfisher.
3. **Single Korean**- Bloom flat with well visible disc. Ray florets are strap like and arranged in 5 or less whorls, *e.g.*, Sharad Bahar, Sharad Shobha, Sharad, Seema, Sharad Singer, Sunset.
4. **Double Korean**- Bloom have visible disc and florets are same as Single Korean class. Number of whorls of ray florets is more then 5, *e.g.*, Jyotsana, Tara, Priya, Sonali.
5. **Decorative**- Developed ray florets make disc invisible. Ray florets are regular or irregular reflexed, *e.g.*, Sharad Mala, Jayanti, Ratna, Neelima and Shabnam.
6. **Pompon**- Ray florets are incurved or reflexed. Plants are short, broad and very systematically and uniformly arranged in hemispherical shape, *e.g.*, Horizon, Apsara, Nanako, Beerbal Sahni and Purple Star.
7. **Sami quilled**- The ray florets are tubular up to certain length of the florets from base and then open at the tips, open tip portion may be flat, reflexed or incurred, *e.g.*, Jean, Alison, Garnet.
8. **Quilled**- Ray florets are elongated and tubular a quilled. The tip of florets may be open but not develop, *e.g.*, Donald, Snow Crystal and Green Night ingle, Reeta.
9. **Stellate**- Florets are like single Korean but both the side of ray florets are reflexed downward disc is flat with short florets, *e.g.*, Stella, Morning Star, Golden Tailor.
10. **Cineraria**- Blooms belonging to flat Korean type with diameter not more than 3cm, *e.g.*, Phyllis, Bindya and Charmis.

Origin and Distribution

Genus chrysanthemum means golden flowers, *i.e.*, *Chrysos* (golden) and *Anthos* (flower). These are native of China, Japan, North Africa and Southern Europe. Chrysanthemum was first cultivated in China as flowering herb and described in ripping an early 15th sanctuary B.C. In India chrysanthemum is successfully grown under protected as well as open conditions in different states.

Varieties

A. Large Flower

1. **White colour**- Snow Ball, Kasturba Gandhi, Innocence.
2. **Yellow colour**- Chandrama, Sonar Bangla Supper Giant, Mountaineer.
3. **Red colour**- Diamond Jubilee, Distinction, and Autumn Blaze.

4. **Mauve colour-** Mahatma Gandhi, Peacock, Classic Beauty, Pink Giant.
5. **Red purple colour-** Tokyo and Gambit.

B. Small Flower for Cut Flowers

1. **White-** Birbal Sahani, Apsara, Himani.
2. **Yellow-** Nanako, Jayanti, Kundan.
3. **Mauve-** Sharad Prabha, Neelima, Ajay.
4. **Red-** Jaya and Zubli.

C. Small Flower for Pot Plant

1. **White-** Sharad Shobha, and Shweta Singer, Rita and Niharika.
2. **Yellow-** Indira, Sonali Tara, Sharad Kanti.
3. **Mauve-** Sharad Prabha, Hemant Singer and Fantasy.
4. **Red-** Rakhi, Arun Singer and Jaya.

D. No Pinch Variety

Arun Singer, Rangoli, Haldighati, Bindiya, Suhag Singer, Mother Teresa.

E. Annual Chrysanthemum Variety

Primrose Gem, Flam Shades, John Bright, Whity.

F. IARI Release Variety

Pusa Guldasta, Pusa Aditya, Pusa Sona, Pusa Anmol, Pusa Chitrakhsa, Pusa Scentanary, Pusa Keshari.

G. NBRI Release Variety

Little Pink, Little Kusum, Little Orange, Himanshu, Khoshoo, Kaul.

Soil and Climate

Chrysanthemum requires sandy loam with pH 6.3-6.7. Chrysanthemum is a shallow rooted crop and is sensitive to water-logging. For cultivation of chrysanthemum EC should may be more than 1-1.5 with low level of salt and sulphate.

The chrysanthemum is a photosensitive crop and it required long day for vegetative growth and short day for flowering. The light and temperature play and important role in vegetative growth and flowering. The best temperature for growing chrysanthemum is range from 20-30°C in day and 15-20°C for night with relative humidity 75-90 per cent.

Propagation

Chrysanthemum can be propagated by both seed and asexual method. In asexual method sucker and terminal cutting is commonly used.

1. **By seed-** For seed propagation seeds are collected after the capatulam is complitaly dry and stored in dry place. After sowing fine compost is used for covering the seed. Regular watering should be done to maintain the proper moisture with the help of water cane. The treatment of Thiourea promotes germination of the dormant seed. Seedlings are ready for transplanting 25-30 days after seed sowing.
2. **By sucker-** Optimum time of separation of plant from mother plant is range 5-6 green leaf appears. Plant cutting from mother plant and then planted in field or pot.
3. **Terminal cutting-** This is commercial method, cutting of 5-7cm length are taken from healthy, disease free plant. Dipping of basal portion of cutting in IBA 1000 ppm for improving rooting percentage.

Land Preparation

Summer ploughing should be done up to 30 cm deep, so that the soil gets exposure to sun and air. If the soil is light sandy stony the next 30 cm deep soil in the ground surface should be returned to the trench with the help of digging fork.

Manures and Fertilizers

Well managed chrysanthemum crop require 50 kg N, 160 kg P_2O_5 and 80 kg K_2O per ha should be provided as basal dose. The half dose of nitrogen, full dose of phosphorus and potash should be applied at final preparation of land. The rest dose of nitrogen should be applied in plants. After one month of transplantation apply liquid fertilizers as 5 g ammonium nitrate along with 5 g potassium nitrate in 10 liter water twice at 15 days interval.

Transplanting

Planting in May-June is ideal for chrysanthemum. In north India planting is done after 2nd week of June. In general purpose it may be planted by the first July or early November. Cutting are transplanted after rooting in June. Plant density of 32 cutting/m^2 is ideal and close spacing at 20-30 cm give more yield than 30 × 30 or 40 × 40 cm space.

Irrigation

Irrigation requirement of plants reduces as it approaches to flowering stage. In field condition generally 3-4 day's interval irrigation is required in summer season. The height and vigour of the plant can be influenced by regulating quality and frequency of irrigation.

Cultural Operation

Weed Control

The pre-emergence application of trifluralin at 4.0kg a.i./ha, reduces the weed population. Manual weeding is beneficial but it is time taking and costly process.

Pinching

Pinching is an important operation for the induce branching and to increase the number of flower in stem. It is three types and required for chrysanthemum plants.

a) **Soft pinching-** This is done to reduce plant size by removing tips of branch. Shoot along with 2-3 open leaf are kept in initial stage.

b) **Hard pinching-** It is done in pot plant. It remove a longer portion to keep the plant compact.

c) **Rollout pinching-** It is practiced only in short plants. Generally two pinching are required–first at 4th week after planting and second at 7th week after planting.

Disbudding

Disbudding is done to remove the side branch, to control flower number and size. It is done in large flowering and decorative types of cultivar. In disbudding auxiliary bud are removed and only terminal but at retained.

Desuckering

During the vegetative growth phase plant growth upward and new suckers continue to develop from the base of the plant. For preventing improper and vigorous growth of plant, suckers are removed from time to time.

De-shooting

It is practices to reduce number of branches for improving the size and form of the flower, *e.g.,* For taking three blooms per plant, three lateral strong shoots are allowed to grow and others are removed at early stage of growth.

Staking

Especially large flowered varieties of chrysanthemum grown outdoors or susceptible to lodging need staking. Earthing up of plants when the spikes start elongating and also provide sufficient support to prevent lodging.

Harvesting

Standard chrysanthemums are harvested at full bloom before central disc begin to elongate. Harvesting is done at early in the morning and harvested stem are placed in water containing solution to keep them turgid. Stem should be cut at least 10 cm above the soil. Woody plant tissue leaf should be removed from lower portion of stems.

Yield

The yield of flower stem varies from 15-20/plant. The average yield of spray type ranges from 1.0 lakh to 1.20 lakh and yield of loose flower ranges from 10-15 tonnes/ha.

Post-Harvest Management

a) **Bunching-** Standard bunch have 10-12 flowers. Spray chrysanthemum 100-200 gm. Flower are wrapped in plastic sleeves and packed in fiberboard boxes.

b) **Grading-** Chrysanthemum are graded on the basis of stem length, flower appearance, flower colour and flower freshness, *e.g.,* For blue colour minimum flower diameter 140mm and stem length 76cm, for red colour minimum flower diameter 121mm and stem length 76cm, for green colour minimum flower diameter 102mm and stem length 61cm.

c) **Storage-** Chrysanthemum are used for long post-harvest life stored dry for 3-4 weeks at 0.5°C. Storage at 2-3°C should not exceed two weeks. Yellowing of leaves can occurs at 5°C in dark but is less like to occurs at 1°C.

Insect Pests

a) **Aphid** (*Myzus persicae*)- This is a sucking type pest. Insect suck the sap and act as a vector for transmitting the various viral diseases. It can be controlled by application of Malathion 0.2 per cent solution or Dimethoate 0.3 per cent or Neem oil 1 per cent.

b) **Hairy caterpillar** (*Diascrisia oblique*)- This is the bitting and chewing type pest. They eat the leaves from the surface. The insect infestation is more in rainy season and it can be controlled by application of Thiodon 35EC at 1.25 ml/lit.

c) **Red spider mite** (*Tetranychus urticae*)- This is the sucking type pest. Their attack is confined to lower surface of the leaves. Insect infestation is more during the month of April-July. It can be controlled by application of Dicofor @ 0.5 per cent, Ethion (0.05) and Wetable sulphure @ 0.3 per cent.

Fungal Diseases

a) **Wilt** (*Fusarium oxyporium* sps. trichothecenes)- This is most severe disease in chrysanthemum cultivation. The symptoms of the disease is chlorosis. In severe case plant wilting takes place. It can be controlled by application of Thiram @ 1.5 gm/lit water as spray or drenching of soil with Carbendazime.

b) **Leaf spot-** Two types of leaf spot are found one is caused by *Septoria chrysanthemella* in which yellow spot appear on the leaves that letter become dark brown and black. The other spot are caused by

Cylindrosporium chrysanthemum in which leaf become dark brown in colour with yellow margin. Disease can be controlled by Carbendazim 0.05 per cent or or mancozeb (0.2 per cent) along with sticker 0.1 per cent at 10 days interval from June-October.

Bacterial Diseases

a) **Bacterial blight** (*Erwinia chrysanthemi*)- This disease can be controlled by disease free planting material. The main symptoms are water soaked reason which finally split the stem. Bacterial infestation is more in high temperature and high moisture condition and it can be controlled by treating the cutting by Streptomycin.

b) **Crown gall** (*Agrobacterium tumefaciens*)- In this problem irregular and round gall appear on the stem and some time leaves. This problem is more common in moist soil condition. It can be controlled by maintaining the appropriate moisture in the soil.

Chapter 28

Gladiolus

- Common name - Sword Lily
- Botanical name - *Gladiolus grandiflorus*
- Family - Iridaceae
- Chromosome No. - 2n = 30 (30-120) X=15
- Origin - South Africa
- Gladiolus named from the Latin word "*Gladius*" means Sword like, because sword like shape of its foliage. The term gladiolus is coined by Pliny the Elder (AD. 23 – 79).
- Leading cut flower of India as well as world.
- 4th rank is international cut flower market.
- Commercially propagated by corms.
- Gladiolus breeder in India are Bajrang Bahadur Singh Bhandari, R.L. Mishra, and D. Mukharji.
- Commercial life of corm is 10- 15 year.
- Flower bud initiations start when the plant is at three leaf stage.
- Most divesting disease wilt or collar Rot (*Fusarium oxysporium gladioli*).

Introduction

Gladiolus is very much liked for its majestic spikes which contain attractive, elegant and delicate florets. These florets open in sequence and longer duration and hence has a good keeping quality of cut spikes. Gladiolus is an import commercial flower crop and is very popular as cut flower, borders, wedding, ports, bouquet and flower arrangement in national and international market. It is a bulbous plant. Gladiolus was introduced into cultivation towards the end of 16th century. In India gladiolus are commercially grown in West Bengal, Maharashtra, Uttar Pradesh, Uttrakhand, Madhya Pradesh, Punjab, Haryana, Himanchal Pradesh, Delhi, *etc.*

Botanical Description

A tender herbaceous perennial gladiolus is grown from seeds and corm. The corm is covered by 4-6 dry scale or husk whose bases are of the older leaf formed during previous growing season. The points where scale is attached to the corm is called node. Each node bears one or two buds. Each bud has a potential to grow into a shoot. The root starts emerging from the base of corm. These roots are called filiform roots. As the plant reached near 4th leaf stage and base of the new corm arises branched called stolen.

Varieties

1. **Pink colour-** Pink Friendship, Pusa Archana, Pink Perfection and Applause.
2. **Green colour-** Lemon Lime, Green Bay, Green Bird, Green Jayat, Green willow.

3. **Cream colour-** Classmate, Cream Topper, Landmark, Ruffled Lotus and Party Ruffles.
4. **Yellow colour-** Folksong, Golden Harvest, Golden Peach, Royal Gold and Lemon Ruffles.
5. **Orange colour-** Tiger Flame, Autumn Globe, Pitter Pears, Ratna Butterfly and Setting Sun.
6. **Red Colour-** Black Prince, Hunting Song, Oscar, Red Beauty, Victoria.
7. **White colour-** Shubhangini, White Prosperity, White Friendship and Snow Princes.
8. **Rose colour-** American Beauty, Royal Brocade, and Upper Trust.
9. **Brown colour-** Brown Beauty, Chocolate Chip, Chocolate Dip and Little Tiger.
10. **Purple and violet colour-** Blue Moon, Mayor, Purple Giant, Pusa Sarang, Pusa Shingarika, Pusa Urmila, Tropic Sea and Rose Delight.

Some Outstanding Cultivar Suitable for Indian Cultivation

Snow Princes, Rose Supreme, Suchitra, Pitter Pear, Mayor, White Prosperty, Pink Friendship, Punjab Dawn, Punjab Morning and Jester.

New Varieties from IARI - Pusa Sindhuri, Pusa Red Valentine, Pusa Shubham, Pusa Kiran, Pusa Vidusi, Pusa Unnati, and Pusa shreejan.

New Varieties from IIHR - Arka Amar, Arka Naveen and Arka Gold.

New varieties from NBRI - Neelima, Urvashi, Suvarna and Roshani.

New varieties from BHU- Malviya Kiran, Malviya Shatabadi and Malviya Kundan.

Soil and Climate

Sandy loam soil is preferred, soil pH is 6-7 ideal, all type are soil is suitable for gladiolus cultivation. When grown on heavy texture soil addition of organic matter or making bed is recommended. Gladiolus plants are sensitive to pollution and salt.

It is a long day plant (12-14 hour). Temperature between 10-25°C is optimum for gladiolus cultivation. Temperature below 10°C arrests growth and development. The favourable low temperature is about 16°C and higher temperature range is 25-30°C. Gladiolus is requiring full exposure to sunlight otherwise blasting may occur or plant may remain blind. Long day temperature delay flowering but improve bloom quality. Storage temperature of corm also affects flowering. Optimum corm storage temperature is 4-5°C.

Propagation

1. **Corms-** It is commercially propagated by corm. For commercially production corm size should be 4-5 cm. Large and medium-sized corm

are used for production of cut spike whereas small sized corm are used as planting stalks for subsequent planting season.

2. **Cormels-** Cormels between 1.0 cm and less than 2.5 cm diameter are grown for production of flowering stalks and increasing the number of good quality planting materials.
3. **Seed**- Seed is used for only breeding purpose.

Land Preparation

One ploughing by cultivator and 2-3 ploughing by harrow followed by planking. Land should be ploughed 1-2 months before at a depth of 20-30cm. Gladiolus cultivated in friable soil. Farm yard manure and other compost are used at the rate 5kg/m^2 and phosphoric and potassium fertilizers should be applied before planting of corms.

Manures and Fertilizers

Whereas the nitrogen should be applied at 300 kg/ha, it is applied in two split dose first at 3 leaf stage and second at 6 leaf stage. Whereas phosphorus and potassium applied as basel dose at 150-200 kg/ha and 120-150 kg/ha at the time of planting corms. Others micronutrient ferrous sulphate applied at 0.2 per cent twice or thrice at 15 days intervals.

Planting

In plains of India, planting time is first fortnight of October. In temperate region corms are planted in March to April. While in place with mild climate like Bangalore, planting can be done all though the year. Corms should be treated with fungicide *i.e.* Carbendazim 0.2 per cent. Generally planting is done in flat bed at the distance of 30×20 cm and 10-15 cm depth for medium corm size 10-60 cm and normally depth of 7cm at this distance about 60000 corms can be accumulated/acer.

Irrigation

Irrigation should be withheld at least 4-6 week before lifting of corm. Normally in sandy soil the crop should be irrigated at 7-10 days interval. A light irrigation just prior to lifting of corm is however beneficial to facilitate their lifting from the soil.

Cultural Operation

Mechanical Weed Control

After planting of corms, the plant requires 4-5 hoeing to keep the weedfree. The manual weed controls are time consumed and increase the input cost.

Chemical Weed Control

It is major common practices for the weed control whereas the use of some pre emergence herbicides are Diuran (0.9 kg/ha), Linuran (3 kg/ha), Alachlor or

Metachlore 4.5 kg/ha and post emergence are 2-4-D (1.5-3.0 kg/ha) Pendamethilin (stamp (pre.) 30 C (1.6 li/ha).

Staking

Especially large flowered varieties of gladiolus grown outdoors or susceptible to lodging as need staking. Earthing up of plants when the spikes start elongating and also provide sufficient support to prevent lodging.

Harvesting and Yield

1. **Flower-** Spikes are ready for harvesting till 60-90 days after planting and continue for about a month for export and distance market. It should be harvest with basal 1-2 floret show colour. For local market, spike are harvested when basal floret is fully open. Harvesting should be done at evening or morning.While cutting spike, 2-3 bottom leaf should be left. Approximately flower spike yield would be around 70000-75000 spike/ acre or 1, 80,000-2, 00,000 spike/ha respectively depend upon planting density, cultivars and management.
2. **Corm –** It generally take 6-8 week after harvesting of spike for the corm to become mature and ready. For harvesting, by manually by Khurpa. After lifting from the soil the upper leafy portion should be remove by twisting and breaking the stock. The cormels should also be separately simultaneously and handled separately. Approximately corm yield would be around 60,000 corm/acre or 2, 00,000-2, 50,000 corm/ha respectively depending on planting density, cultivars and corm size.

Post-Harvest Management

1. **Flower-** The spike are graded before marketing. Spike length 107cm, number of florets 16-18 (minimum) 14,12,10 Graded spike are made into bunch of 10-12 and loosely tide with rubber band and packed in card board boxes. For export purpose spike need to be pre-cooled at 4-5°C for spike 72 hours under dry condition un-cooled spike can be transported for 24-48 hours at 20-25°C.
2. **Holding solution-** Sucrose (4 per cent) + Ammonium sulphate (300 ppm) or Sodium hydrochloride (50 ppm) or both can be use for opening of floret and improving the vase life of spike. By use of the 8HQC (Hydroxy Quinoline Citrate) cut spick can be stored at 1-2°C for 3-4 month.
3. **Corm storage-** The storage of corms at the very low temperature (4-5°C) is an established commercial practice.

Insect Pests

1. **Aphid** (*Aphid gossypii* or *Dyseplus tulipae*)- There are two kind of females– the wingless aphid and winged aphid. The wingless aphid varies from pale yellow to very dark green and produce young aphids while the winged aphids which are seldom seen on the plant. The aphid is transmitting agent of mosaic viruses and it can be controlled by spraying of Metasistox or Sevin or Molathion (0.2 per cent) at fortnightly interval.
2. **Thrips** (*Taeniothrips simplex*)- The thrips are mainly infested the leaves and flowers of the gladiolus plant. The young thrips are light yellow in colour and move about irregularly when disturb. It can be controlled by spraying of Diazimon, Cygon, Methoxychlor or Sevin (0.2 per cent) before flowering.
3. **Borer** (*Helicoverpa armigera*)- Larvae is the main damaging stage of borus and it feeds on leaves and unopen floret. It can be controlled by application of Thiodan 35C (0.5-0.8 per cent) at first instant of larvae.

Nematode

Root knot (*Meloidogyne incognita*)- The affected plant show retarded growth and turn pale yellow in colour and the characteristic symptoms is formation of gall on roots of the plant. It can be controlled by application of Thimate 10 G or Temik 10 G @12 kg granules per acre.

Fungal Diseases

1. **Fusarium wilt-** (*Fusarium oxysporum* var. *gladioli*)- This is the soil born fungus. Its enters in the plant through corm. In severe case leaves become pale yellow and sickle shape. It can be controlled by applying suitable crop rotation, sanitation, curing and early lifting of corm at 29-30°C about 1 week together. Some tolerant varieties are Sylvia, Dheeraj, White Friendship, Apricot Globe, Suchitra, *etc.* and can be controlled by pre-planting or pre-storage treatment of corm Carbendazim (0.1 per cent) or Mancozeb (0.1 per cent).
2. **Dry or neck rot** (*Stomatinia gladioli*)- This disease is also known as root rot and it is characterized by small, dark superficial spot on the storage corm. It can be controlled by application of Blitox (0.Blitox (0.3 per cent) 2-3 times during growing season at fortnightly interval.
3. **Storage rot** (*Penicilium gladioli*)- This disease is mainly appear in the storage chamber. In severe cases corms become black, brown or greenish due to development of mold. It occurs only due to poor air circulation which results in emission of a foul smell. This problem can be controlled by damp storage and high temperature more than 5°C.
4. **Flower rot and botrytis blight** (*Bortytis gladiolus timmermanns*)- The main symptomps of this disease is that brown patches appear in the tissue,

watery spot on the flowers and in initial stage leaves are yellow in colour and later on it becomes dark brown round spot. It can be controlled by regular application of Maneb, Benlate, Vinclozolin from the times the crop comes up, during the weather at least weekly @ 0.2 per cent should be done.

Bacterial Disease

1. **Bacterial leaf spot** (*Pseudomonas marginata*)- This disease is more common in dense population of the plant. The field sanitation is helpful in reducing the spred of bacteria and it can be minimized by diping the corm in Mercuric chloride 1:1000 before planting and storage.

Physiological Disorders

1. **Tip burn-** In this disorder discoloration and drying up of tips of leaves due to high level of fluorides in the atmosphere. Application of Blitox 50WP (0.3 per cent) can minimize the problem.
2. **Geotropic bending of spikes-** The bending of spike is primarily due to the lateral downward movement of auxin, indole 3 acetic acid and its accumulation on the lower portion of the spike. The tips of gladiolus spikes show tendency to bend against gravity if placed horizontally for longer period. It can be controlled by application of IAA 25-50ppm.

Chapter 29

Marigold

- ☆ Common name – Gainda
- ☆ Botanical name – *Tagetes* species
- ☆ Family – Compositae
- ☆ **African marigold -** *Tagetes erecta*
- ☆ Chromosome No. - 2n=24 (X=6)
- ☆ Origin – Mexico
- ☆ Length - Tall (90cm)
- ☆ Duration - 2-2.5 month.
- ☆ **French marigold -** *Tagetes petula*
- ☆ Chromosome no. - 2n= 48 (X=6)
- ☆ Origin - Mexico and South America
- ☆ Length - Dwarf (30-40cm)
- ☆ Duration - 2.5-5 month
- ☆ **Others –**Single signet – *Tagetes tenuifolia*
- ☆ Sweet scented - *Tagetes lucida*
- ☆ African marigold is also known as "Rose of Indies"
- ☆ Highest essential oil content is found in *Tagetes signata* suitable for perfumes industry.
- ☆ Wild marigold *Tagetes minuta* suitable for essential oil for perfumes and cosmetics.
- ☆ Highest content of Zea xanthin and Xanthophylls pigment in African marigold. The dried petals are used in poultry to increase the colour of egg yolk.
- ☆ Xanthophylls (Lutein) 82-90 per cent are the major constituents in the flower petals.
- ☆ French marigold is a yellow tetraploid- *Tagetes erecta* x *Tagetes tenuifolia.*
- ☆ Interspecific hybrid between African marigold and French marigold is developed by America.
- ☆ Marigold is a cross pollinated crop.
- ☆ Pinching is done after 40 day of planting.
- ☆ Volatile oil content of French marigold is 0.5- 1.5 per cent.

Introduction

In India marigold is one of the most commonly grown flowers and used extensively on religion and social function in different forms. It was introduced in India during the 16th century and since then it has been naturalized in different Agro-climatic region of India. Marigold is an important flower crop in India. They are extensively used for making garland, beautification and other purpose, *i.e.,* oil extraction and therapeutic use. Marigold plantation has been found the beneficial to reduce the nematode. Root knot (*Meloidogyne* spp.) can be effectively managed in the field without pesticide.

Origin and Distribution

The name "Tagetes" was given after "Tages" a demigod known for his beauty. Neither African marigold nor French marigold comes from Africa or France. It is native to central and southern America, specially Mexico. During early 16th century it spreads to various part of the world form Mexico. In India marigold was introduced by Portuguese and it became popular and spread quickly because of its wide adaptability to varying soil and climatic condition.

Varieties

1. African Marigold

Alaska, Apricot, Golden Age, Crown of Gold, Fire Glow, Giant Double African Orange, Gold Smith, Happiness, Happy Face, Hawaii, Honey Comb, Sugar and Spice, Texas and Yellow Climax, Sunset (plant are 90-120 cm tall, large double flower, bloom in beautiful shade of yellow and orange and flowers are long lasting).

Descriptions of some important varieties of African marigold are given below.

- ☆ **Pusa Narangi Gainda** (Cracker Jack × Gloden Jubiliee)- It takes 125-136 days for flowering after sowing, plant height 73.3cm, orange colour, carnation type compact, 7.8 cm in size, disc floret are presents and yield 25-30 tonnes/ha.
- ☆ **Pusa Basanti Gainda** (Golden Yellow × Sun Giant)- It takes 135-145 days for flowering, duration 40-45 days, height 58.8cm, vigorous, uniform, carnation type, double, compact, 6.9 cm in size, disc floret present but invisible, yield 58 flower/plant.
- ☆ **Pusa Arpita-** Orange colour, medium size, flowering during December-February in north. Flower are compact with turmeric colour, average yield is 18-20 tonnes/ha.
- ☆ **Hisar Beauty-** It is released from Haryana Agriculture University, Haryana, dwarf and compact flower dark red petals with yellow margin, suitable for exhibition, wedding and potting, average yield is 40-50 flower/plant.
- ☆ **Arka Bangara-** It is released from IIHR Bangalore. Flowers are yellow, gold colour and average yield is 10 tonnes/ha.
- ☆ **F_1 Hybrid varieties-** Apollo, Climax (First F_1 hybrid) First Lady, Orange Lady, Moon Shot and Show Boat (3x sterile).
- ☆ **Nugget-** Interspecific triploid hybrid develops from USA.
- ☆ **Sunset-** Its flowers are tall and large, double flower bloom is beautiful shades of orange and yellow.

2. French Marigold

Descriptions of some important varieties of French marigold are given below.

a) **Single type-** Naughty Marietta, Sunny, Legion of Honour.

b) **Double type-** Bovita, Bolero, Butterscotch, Zipsy, Harmony, Orange Flame, Rusty Red, Red Brocade, Spray, Spun Gold, Spun Yellow, Melody, Spray, Lemon Drop, Goldie.

Some other varieties of French Marigold are given below.

Disco Red- It is the annual with single flower bright red, yellow, center. Flower size is 3 - 5 cm, upright 40 cm tall and plant spread is perfect for container.

Marvelous Orange- The longest flower fresh crested marigold in the market, fully crested bloom, often measure up to 7 cm in diameter and flower height is 30 cm for early flowering.

Lemon Drop- Short 12-20 cm plant, producing abundance of cannery yellow flower that looks great in pot.

Bonanza Blend- It is large flower having bright orange sunshine yellow vibrant bicolour flower, height is 20-30 cm and plants are drought tolerant to some extent.

Soil and Climate

Marigold can be grown successfully in a wide range of soil. Soil should also be well drained and well fertile soil, pH should be 7 - 7.5 but it can be grown under slightly acidic soil with pH 5.8-6.2.

Marigold requires mild-climate of luxuriant growth and profuse flowering. African and French marigold both are hardy in nature. It can be grown throughout the year under tropical and subtropical region. Mild climate during growing period 14-28°C greatly improve flowering while higher temperature (26-36°C) adversely affects the flower. Night temperature 15-18°C and day temperature 18 - 22°C is ideal for producing high yielding plants. The critical photoperiod for *Tagetes erecta* is 12.5 and 13 hour.

Propagation

1. **By seed-** The marigold seed is black in colour and count about 300-350 seed/gram. The seed remains viable for about 1-2 year. Almost all the cultivar of African and French marigold can be propagated by seed. The rate of the *Tagetls patula* is 1.5-2.00 kg/ha and *Tagetls erecta* is 1-1.5 kg/ha for transplanting. About 8-10 seed bed of 3 x 1 m^2 size are needed to raise seedling for 1 ha. Seed should be sown at space 6-8 cm and depth 2 cm then covered with the layer of the soil sand, leaf mould and FYM and watering is done daily with watering cane. The seed germination occurs in 5-10 days of sowing at temperature 18-30°C. For summer season seeds are sown in the second week of February while crop of rainy season, sowing

of seed is done in the first week of June and in winter season crop seed is sown 2nd week of September.

2. **By cutting-** Plant can be easily multiplied during rainy season by cutting. During this period emergence of best root from stem also help for the good stabilization of cutting. Apical portion of the shoot is found suitable for cutting. Cutting should be about 6-8 cm long and treated by IBA @ 500- 1000 ppm to enhance the root.

Land Preparation

For preparation of land ploughing is done 3-4 times and levelling is done after the ploughing. Ploughing at 30cm depth is done nearly a fortnight prior to transplanting of seedlings. During the last ploughing application of well decomposed farm yard manure at 20 tonnes/ha is done. Good drainage facility especially for rainy season crop is necessary.

Manures and Fertilizers

The French and African marigold both are well responded to chemical fertilizer. In addition to farm yard manure 20-25 tonnes/ha application of 200: 80: 80 kg/ha NPK is good for flower yield. Full dose of phosphorus and potash and half dose of nitrogen should be applied at the time of field preparation. Rest dose of nitrogen should be applied in two split dosage after 30 and 60 days of transplanting.

Planting

The age of seedling should be around 30 day for transplantation having 3-4 leaf. Seedling should be transplanted in well prepared land and after transplantation light irrigation is essential. In general a spacing of 40×40 cm for African marigold and 30×30 cm for French marigold is recommended.

Irrigation

In marigold irrigation depends upon the soil type, and climatic condition. Summer crop should be irrigated at 4-5 days interval which in rainy season crop should irrigated as per atmospheric condition. During heavy rainfall drainage is essential for removal of excess water from the field. In winter crop light irrigation should be done at 8-10 days interval. First irrigation should be done just after transplantation.

Cultural Operation

1. **Weed control-** In marigold cultivation weed control is very important operation. The application of Pendimetheline @ 2lt a.i./ha should be beneficial at pre-emergence in French marigold. Generally 3-4 manual weeding are required for entire growth period.
2. **Hoeing-** To create conductive soil condition for plant growth, hoeing should be done at short interval to keep the soil porous and enable light, air and water to reach the root easily. It improves the moisture retention capacity and help to remove weeds.

3. **Pinching-** Removal of apical portion of plant is called pinching. Marigold plants grow straight upward to their final and develop into terminal flower bud. If the terminal portion of shoot is removed early, emergence of side branches occur which results more number of flowers. Pinching is done 1st at 40 days after transplantation and 2nd at 60 days after transplantation for higher yield.

Harvesting

Flower in winter crop starts by middle of June and continues till March. In summer season crop flowering, in middle of May with maximum intensity in the month of June to onset of rain. Plucking of flower is done in the evening or morning and regular plucking of flower enhances yield considerably. After harvesting flower should be collected in bamboo basket for carrying to the market. For local market marigold flowers are taken into gunny bags whereas for distant market bamboo basket are used.

Yield

On an average fresh flower yield is 12-15 tonnes/ha during rainy season, 15-18 tonnes/ha in winter, 12-14 ton/ha in summer season obtained. In general yield of 20-22 tonnes flower/ha are found in the African marigold and 12-15 tonnes/ha can be obtained in French marigold. On an average oil yield is 50-60 kg/ha.

Insect Pests

1. **Spider mite** (*Tetranycus* spp.)- Spider mites appear more at the time of flowering period. Plant appearance is near to dusty and it can be controlled by Kelthane (2ml/lit).
2. **Hairy caterpillar** (*Diacrisia oblique*)- This insect is polyphagus in nature and it damages the foliage of the plant. This pest can be controlled by Indosulphon (2ml/lit.).
3. **Leafhopper** (*Empoaea fabae*)- This is the more common insect in marigold. The symptoms of infestation are cupping or rolling of leaves, wilting of shoot tips and leaflets. It can be controlled by Malathion or Rogar (2ml/lit.)

Fungal Diseases

1. **Damping off** (*Rhizoctonia solani*)- This is the soilborne disease and in severe case disease appear as brown spot girdling and later on pre-mature mortality. It can be controlled by sterilization of soil before raising the seedling and post emergence seedling treatment with copper fungicide like Blitox (0.4 per cent).
2. **Collar rot** (*Sclerotium rolfsii*)- This disease is caused by soilborne fungus. This disease is more common in nursery stage. The drenching of soil

with Carbendazim and Metalaxyl at 0.2 per cent solution minimize the problem.

3. **Leaf spot** (*Alternaria alternate*)- Leaf spot mainly appear in leaves and stem as purplish spot. Application of Bliox 0.1 per cent and Bovistin 0.1 per cent effectively control disease.
4. **Fusarium wilt** (*Fusarium oxysporium*)- The causal agent is soil born. Seedlings are killed, in older plants black streak, darken the vascular tissue up to one side of the plant. During wet weather disease infestation is more and it can be controlled by application of Carbendazim 0.2 per cent.

Bacterial Disease

Bacterial leaf spot (*Pseudomonas cinerea*)- In this disease 2-5 mm small circular spot are formed on leaves and petiol. This disease can be controlled by destroying the infected plant and overhead irrigation.

Physiological Disorder

Leaf burn- The tips and margin of leaves turn yellow and die due to excess of Boron, Molybdenum or Manganese deficiency. The application of micro-nutrient is not more than 55 ppm, 3 ppm and 24 ppm of Manganese, Boron and Molybdenum respectively is beneficial.

Chapter 30

Anthurium

- Botanical name - *Anthurium species*
  (*Anthurium andreanum, Anthurium brown*).
- Family - Araceae
- Chromosome No. - 2n= 30 (X=15)
- Origin - Colombia
- The term anthurium is derived from Greek word *Anthos* means flower and *Oura* means tail, referring to spadix.
- It is the second cut flower among the tropical cut flower plant, major exporter of anthurium is Hawaii.
- Flowers are protogynus in nature.
- Inflorescence type is spadix.
- Pollination type is cross pollination.
- Fruit type is Berry.
- Economic part is spathe.
- **Spathe** - The flower consists of a colourful modified leaf and spathe and growth is double sigmoid growth curve. Most preferred colour of the spathe is red followed by pink.
- Anthurium bloom throughout the year, one bloom arises from the axils of the each leaf.

Introduction

Anthurium is a perennial, herbaceous plant of great beauty and grown either for its show cut flower or for attractive foliage. It is also known as flamingo flower or flamingo lily. In India the cut flowers of *Anthurium andreanum* are cultivated for considerable length of time. It is commercially cultivated in North East state and Kerala. The development of main flower distribution center in Dubai offers important possibilities for Indian floriculture.

Genetics

Anthurium andreanum, Anthurium brownie, Anthurium crystalenlum (2n= 30), *Anthurium magnificum* (2n = 32) diploids, *Anthurium scandens* (2n = 45) are triploid and *Anthurium digitatum* (2n = 60) are tetraploid.

Botanical Description

It is semi-terrestrial and perennial plant with crippling to arborscent stem. Leaves are large carried on long petioles and have a heart shaped appears. The bisexual flowers are small and densely packed on a cylindrical spadix subtended in large heart shaped spathe.

Varieties

1. **Red colour spathe-** Mauritius Red, Nova, Temptation, Sweet Heart, Scarlet Red, Flame, Cherry Red.
   a) **Haland -** Tropical Red, Scarlet, Flaking, Jertrood, Cancan Fla Red.
   b) **Hawaii -** Kosohara, Koumana, Hayashi, Mickey Mouse.
2. **Orange colour spathe –** Mauritius Orange, Nitta, Peach, Sunshine Orange.
   a) **Haland –** Fla Orange, Casino.
   b) **Hawaii -** Nitta, Sunburst, Diamond Jubilee.
3. **Pink colour spathe –** Agnihotri, Passion, Paradise Pink.
   a) **Halland -** Honette, Surprise, Spirit, Cheers, Sonata, Magic Pink.
   b) **Hawaii** - Candy Stripe, Candy Queen, Abepink.
4. **White colour spathe** - Mauritius White, Lima White, Haga White, Linda Dee Mol.
   a) **Halland -** Cuba, Gousha, Bienco, Uranus.
   b) **Hawaii-** Manoa Mist, Chameleon, Morocco, Suchiro.

Soil and Climate

Anthurium requires well aerated, medium textured and good water holding capacity soil. The optimum range of soil pH or media is from 5.5 – 6.0. Soil should be rich in organic matter. Leaf mould, cattle manure, coconut husk, tile pieces, sand and coir dust may be used to prepare a suitable medium.

The anthurium best day temperature range from 25-28°C and night temperature 18-20°C. Temperature above 32°C may cause foliar burning, faded flower colour and reduce flower light. The optimum temperature for flower development is 20°C and relative humidity 80 per cent is ideal.

Propagation

1. **By Cutting**- Terminal cutting of anthurium are generally used for propagation. Intermitted mist has been found more effective in term of more rooting percentage. Treatment of terminal cutting with 500 ppm IBA produce longest flower stalk with more diameter of spathe.
2. **By Division-** It can easily multiply by division of offshoot with portion of aerial root from the main stem. Generally anthurium cultivars produce 14-16 offshoot/plant/year.
3. **By Micro-propagation-** Anthurium are multiplied very rapidly through tissue culture and produce more flower than those plant raised from seedling. Leaf segment, vegetative bud, spadix segment, stem section, *etc.* have been successfully used as explants.
4. **By seeds-** Growing of anthurium plant from the seed is lengthy and flowering is attained 2-3 year after germination.

Land Preparation

Bed culture is recommended for commercial plantation rather than growing them in pots. Thus, flat land is preferred for the construction of beds. However, if land is sloppy, terracing should be recommended prior to preparing beds.

Beds should be dug 60-90 cm deep and the bottom overlaid with medium size pieces of tile or charcoal to enhance drainage. The beds are then filled with the growing media consisting of leaf mould, cattle manure and sand mixed together as explained (in soil) above. Every year this media should be topped up either with half decayed leaf mould or the same mixture.

Manures and Fertilizers

Anthurium requires split application NPK at 30:20:50 gm/m^2/year at 3, 6 and 9 month after planting. Increased number of suckers, stalk length, length and width of spathe, number of flower/plant, improve flower weight and give early flowering. Fertilizers application should be done by the means of foliar application upto the age of 4 month. The ratio of NPK is 30:10:10 kg/ha may be spread at 0.05 per cent once a week. When plants are one year old application of 1 per cent solution of NPK at the ratio of 17:17:17 at fortnightly interval give better result.

Planting

Generally 8-10 month-old plants are planted at the rate of 65000 plants/ha. The spacing should be maintained at 30 × 30 cm row plant. The closer planting would require the constant maintenance of 3-4 leaves/plant at all time.

Irrigation

Plants should be irrigated at least twice daily. During peak summer 3-4 times spraying of water may be required. Mist/overhead sprinkler irrigation is the best for anthurium. Irrigation in the afternoon should be timed in such a way as to leave sufficient time for water to evaporate so that plants do not be dump during night hours.

Cultural Operation

Weed Control

Weed control is the very essential operation in anthurium cultivation. The pre-emergence application of Lontral 3-EC (Clopyralid) at 0.28 kg/ha should be applied for weed control. The constant manual weeding and earthing minimize the weed population.

Shading

Shading from sunlight is necessary for normal growth. Shade requirement usually range from 50-90 per cent are sunlight. A minimum height 3 m from the ground is required in tropical plains to facilitate ventilation. Cultivar, age of the plant and climate under which it is grown are the main shading factor.

Removal of Suckers

Removal of the suckers is very important operation of anthurium cultivation. The unwanted suckers should be removed to prevent the withdraw of extra energy from the parental plant. That is why plants are straight and distributions of food material in desired plant part resulting big flower size.

Harvesting and Yield

Anthurium flowers are generally harvested when spadix is almost fully developed. Flower picked to early wilt quickly. The best time for harvesting is when one-third to two-third of the true flower on the spadix are open. Anthurium flowers always harvested in the morning with help of a sharp knife. After harvesting flowers are immediately shifted in the water to prevent drying up of flower.

Generally one plant produces 2-3 flower/year. Which increases 4-6, 6-8 flower/ plant during second and third year respectively. Each plant also produces 3-5 and 6-8 propagating material during second year and third year of plantation.

Insect Pests

1. **Aphid** (*Myzus persicae*)- This is the sucking type pest. They suck the sap from flowers, and leaves. It can be controlled by systemic insecticides like dimethoate at 2 ml/lit of water.
2. **Mealy bug**- This is the most serious problem of anthurium. Generally mealy bug appear on the lower surface of leaves and the roots. It can be controlled by drenching of soil with dimethoate (0.2 per cent).
3. **Mites**- Spider mites are most severe in summer month. It is very small, oval shaped, spider like, white green transparent insect. It can be controlled by spray of Dicofol (18.5 EC) 2.5 ml/1it or Triazophos (40EC) l.5 ml/lit at 15 days interval followed by pongamia or neem oil at 10 ml/ lit.
4. **Thrips**- Thrips also suck the sap from leaves and causes mottling of the leaves and plant. It can be controlled by Kelthane 0.1 per cent.

Fungal Diseases

1. **Anthracnose** (*Colletotrichum gleosporioides*)- It is also known as black nose or spadix rot. This fungus primarily infects individual flowers in the spadix. This fungus can also infect leaves, produced elongated diamond shaped lesions. It can be controlled by Dithane -M-45, Carbendazime @ 1gm/lt. However, badly affected plant parts should be removed and destroyed.
2. **Powdery mildew** (*Erysiphe communis*)- Powdery mildew is the most common disease in anthurium. The high temperature and high moisture condition are the most favourable for growth of the fungus. White powdery substances appear in aerial part of the plant and it can be controlled by benomyle (0.1 per cent).

3. **Bacterial blight-** (*Xanthomonas campestris* sp. dieffenbachia)- The main symptoms of this disease are small, scattered, angular water soaked spot near the margin of leaves and yellowing of the whole plant. The application of ammoniacal fertilizer should be avoided and it can be controlled by spraying streptomycin sulphate (200mg/lt) at weekly interval.

Physiological Disorders

1. **Folder ears -** In this disorder the basal portion of the flower is not fully unfolded. This disorder is mostly seen at the time of harvesting of flower. This is a genetical problems and exact remedy is not known.
2. **Sticking -** Flower does not open because the spathe is stuck as a result of sticking. This disorder depends upon the variety and during rapid growth period. Low relative humidity is seen to have a negative effect.
3. **Jamming -** Jamming is mainly variety dependent characteristics but more frequently occur under arid condition. The flower jamms in the sheath since this leaves is wound very tightly around the flower. It can be minimized by regular and light irrigation.
4. **Cracks -** This disorder is mainly due to higher relative humidity. In this disorder the both side of the spathe are damaged in high humidity condition. Lower humidity during the night is effective to reduce crack in anthurium.

Chapter 31

Jasmine

- ☆ Common name - Chameli
- ☆ Botanical name - *Jasminum species*
- ☆ Family - Oleaceae
- ☆ Chromosome no. - 2n =26 (X=13)
- ☆ Origin - India
- ☆ The word jasmine comes from Arabic word Yasemin.
- ☆ Jasmine is the perennial plant
- ☆ Origin of the Arabian jasmine East Indies or India and mullai of South India
    1. **Arabian jasmine** - *Jasminum symbac,* pruning time April, spacing, 1.2 × 1.2 m.
    2. **Mullai jasmine** - *Jasminum auricultra,* pruning time December - January, spacing - 1.8 × 1.8 m.
    3. **Royal Spanish jasmine** - *Jasminum grandiflorm,* pruning time Mid December, spacing 1.5 × 1.5 or 1.8 × 1.8.m.
    4. **Italian jasmine**-*Jasminum humile,* yellow colour, flowering time April to June.
    5. **Tree jasmine** - (Shrubbery tree) *Jasminum arborescence,* flowering time November-May.
    6. **White Jasmine**-*Jasminum officinalis,* flowering time spring.
- ☆ Double flowering types of *Jasminum sambac* are known as Motia or Mogra.
- ☆ Most successful method of layering in Jasmine is ground layering.
- ☆ World famous Jasmine oil extraction, from Spanish Jasmine.

Introduction

Plants are grown both as shrub and climber. Flowers and flower buds are used for making garland and veni and for religious offering. Jasmine oil extracted from the flower is high valuable and used in manufacturing perfumes, cosmetics and hair oil, *etc.* Flowers are used in Jasmine tea or herbal or black tea. In India it is grown in different state especially Tamil Nadu, Kerala, Karnataka, Uttar Pradesh, Bihar and West Bengal.

Genetics

The basic chromosome number of jasmine generally diploid is n = 13 and 2n = 26. However higher ploidy level was also observed in Gundumali variety (*Jasminum sambac*) (2n = 39) and *Jasminum angustifolium* (2n = 4× = 52).

Origin and Distribution

Jasmines are native of tropical and subtropical region. The Arabian Jasmine is considered as native of East Indies and its original home being the region of West

Indies. The *Jasminum auriculatum* is distributed in western peninsula of India and it's a native of south India and the central provinces. Jasmine are preferably grown on commercial scale in various part of India particularly, Coimbatore, Madurai, Athur, Chennai, Kalupathi, Tamil Nadu, Bengal, Mysore. In Uttar Pradesh jasmine is grown in Kannauj, Jaunpur, Varanasi, Allahabad and Farukhabad.

Varieties

1. *Jasminum auriculatum* – Co-1, Co-2 and Parimulai.
2. *Jasminum grandiflorum* – Co-1, Co-2 and Arka Surbhi.
3. *Jasminum sumbac*- Double Mogra, Gandumali, Kasthurimali Ramabanam, Suji-mali, Khoya, Aradhana, Single Mogra, Belajapani, Arka Aradhana.

Soil and Climate

Jasmine requires clay soil for vegetative growth but flower production is more in gravel soil. Jasmine performs well in well drained, rich in organic matter and sufficient water drainage. The optimum soil pH vary from 5-8. Clay soil can be improved by aeration of lime and application of organic manure.

For proper growth and flower production mild tropical climate is good. Suitable temperature range between 11- 27°C for proper growth and flowering. For proper growth and flowering jasmine require 13°C in night temperature and 18°C in day temperature.

Propagation

1. **Cutting –** Jasmine is commercially propagated by cutting. For better rooting coarse sand should be used. The best time for cutting is April to September. *Jasminum auriculatum, Jasminum grandiflorum, Jasminum sambac* stem cutting performe well after the treatment of IBA 4000 ppm.
2. **Layering-** Jasmine is also propagated by simple layering but it take more time for success and limited number of plant can be prepared from a bush. The best time for simple layering is rainy season or autumn season.
3. **Budding-** *Jasminum auriculatum, Jasminum grandiflorum, Jasminum sambac, Jasminum aungustifolium and Jasminum communis* are commercially propagated by patch budding. The best time for budding is in the month of July-September.
4. **Seed-** Seed propagation is generally used for breeding purpose. *Jasminum* sps. like *Jasminum auriculatum, Jasminum angustifolium, Jasminum grandiflorum, Jasminum pubescence, Jasminum regidum, etc.* have been reported to set the seed.

Land Preparation

Jasmine is a perennial crop and it performs well in well pulverized soil of good depth free from stone and other hard pan. Summer ploughing is essential for good crop. The provision of proper water drainage and irrigation channel should be provided after ploughing. Pits of 45 cm depth and diameter are dug out 30 days before planting and exposed to sun. The pits are filled with two part of farm yard manure, one part of coarse sand and one past of soil at the time of planting.

Manures and Fertilizers

The balance nutrition is required for proper growth and flower production. The application of NPK should be done @ 60:120:129 gm/plant/year with 10 kg FYM. The application of fertilizer in two split doses–first dose should be given in June-July and second dose should be given in the month of January.

Irrigation

Jasmine requires regular supply of water during spring and summer. The scarcity of water during these months will hamper the growth and flowering drastically. In sandy and sandy loam soil irrigation should be done at an interval of 3-4 days. Whereas, for heavy textured soil frequency of irrigation can be delayed to 10-15 days. During the winter month irrigation can be done at the interval of 10-15 days.

Cultural Operation

Pruning

Pruning is most essential operation in jasmine cultivation. The flowers of jasmine emerges always in current season growth. For promoting the new shoot, pruning should be done regularly. In general pruning should be done in late winter season for removing the all dead dry branches and cutting back past season branches up to 45cm from the ground level. In north Indian condition the best time for pruning is the end of January.

Harvesting

Jasmine starts flowering from second-year of planting but economic yield obtained after third-year of planting. Fully developed unopened fresh flower are picked early in the morning, why for extraction of concreate only fully opened flowers are required.

Yield

Yield of well managed jasmine crop are as follows:

Jasminum auriculatum - Flower yield 4636 - 9022 kg/ha and concrete yield - 13.44 - 28.24 kg/ha.

Jasminum grandiflorum - Flower yield 4259 – 10144 kg/ha and concrete yield - 13.85-29.29 kg/ha.

Jasminum sambac - Flower yield 2063 - 8129 kg/ha and concrete yield 11.28-15.44 kg/ha.

Insect Pests

1. **Bud worm –** (*Hendecasis duplifasciallis*)- The bud worm is very serious pest in jasmine cultivation. Larva is greenish in colour with a black head bores into immature buds and feeds on floral structures. It can be controlled by application of Monocrotophos 2 ml/lit.
2. **Mite-** (*Steneotarsonemus pallidus*)- The warm and dry weather is most favourable for the incidence of red spider mite. The mites are generally appear in lower surface of the leaves and suck the sap and plants become yellow and drop off. Application of Thimet 2g/plant is beneficial and it can be controlled by application of Metacid 0.15 per cent.
3. **Citrus whitefly-** (*Dialeurodes citri*)- Citrus whitefly is also infect the crop but there is no economic loss has been reported. It can be controlled by application of Malathion 0.15 per cent or synthetic pyrethroids.
4. **Nematode-** (*Meloidogyne incognita*)- Nematode can be controlled by application of Memagon @ 4 g/plant.

Diseases

1. **Leaf blight** (*Cercospora jasminicola*)- Leaf blight is a very serious disease in jasmine. It appears on the upper surface of the leaves and spread rapidly during the rainy season. Flower production is adversely affected during heavy infestation. It can be controlled by application of Benlate- 0.2 per cent, Dithane-M-45- 0.1 per cent and Bavistin-0.15 per cent.
2. **Rust** (*Uromyces hobsoni*)- This disease is mainly appeard in July-August during rain. Blisters are appearing causing yellowing of leaves. It can be controlled by dusting of Sulphur 15-20 kg/ha or application of Copper oxycloride 0.15 per cent.
3. **Wilt** (*Fusarium solani*)- Wilt is a soilborne disease. Symptoms appear on the lower leaves and spread gradually to the whole plant. It can be controlled by drenching of soil with Bavistin 0.1 per cent solution.

Chapter 32

Carnation

- ☆ Common name - Divine flower, Garden pink
- ☆ Botanical name - *Dianthus caryophyllus*
  - *Dianthus barbatyus,*
  - *Dianthus chinensis.*
- ☆ Family - Caryophyllaceae
- ☆ Chromosome number - 2n = 30 (X=15)
- ☆ Origin - Southern France
- ☆ The term Caryophyllus is derived from Greek word *Caryon* means Nut and *Phyllon* means Leaf.
- ☆ The term Dianthus is Greek word means divine flower. This is a cool season crop and herbaceous, half hardy perennial flowering plant quantitatively long-day plant.
- ☆ First introduced carnation in India is sim type (1980). More demand in Indian market is of standard carnation (one large flower per stem).
- ☆ A pigment in flower is due to Acynic group-white yellow, Cyanic group-red, lavender, pink and Transition group-crimson colour.
- ☆ First interspecific hybrids develop in world by "**Thomas fair child**" in 1717 crosses between Carnation x Sweet William.
- ☆ Pioneer carnation breeder "**Montage allowood**".
- ☆ Popular carnation type in India - Clove scented marguerite carnation.
- ☆ Singleness flower is controlled by incomplete recessive gene doubling, controlled by monogenic dominant gene.
- ☆ Annual carnation is propagation by perennial terminal cutting.
- ☆ Pinching or stopping is an important operation in the successful production of carnation.
- ☆ Tinting means application of artificial colour in carnation flower.

Introduction

Carnation is one of the most beautiful and commercially important cut flower crop. Carnation flower can easily withstand long-distance transportation as it has and excellent keeping. The first genetically modified carnation variety "Moon Dust" was released in Australia by Florigene Company and other transgenic variety was that of "Moonshadow". The major breakthrough in carnation flower industry was evolvement of cultivar "William Shim" (in 1938-1939) by "William Shim" of the U.S.A. In India carnation is grown intensely in Nasik, Pune, Coimbatore, Bangalore, Delhi, Kolkata, Uttrakhand and Shimla.

Botanical Description

The yellow colour in carnation is believed to have been contributed by *Dianthus knappi*. Modern-day perpetual flowering carnation is *Dianthus chinensis* into *Dianthus caryophyllus.* Other species like *Dianthus arenarius, Dianthus barbatus, Dianthus plumarius, Dianthus winteri, Dianthus noblis* were also utilized for the development of modern carnation cultivar.

Classification of Carnation

1. **Perpetual flowering** - These are hybrid of different *Dianthus species* and grouped into two class that is standard and spray. They bear flower around the year and most suitable for cut flower.
2. **Chaband/Margarita**- These are annual carnation, propagated by seed and flower are single or double. This type is developed by crossing of *Dianthus chinensis × Dianthus caryophyllus.* It is found various type, *i.e.,* Giant chaboud, Compact dwarf chabound, Margarita.
3. **Border and Picotee** - The flowers of border type carnation are symmetrical, in which picotee type ground colour is without spots or bars. The edges are regular and of decided bright colour, not dashing in ground colour of rest of the petal. Border type carnations are classes according to the colour of flower, *i.e.,* Flakes, Selfs, Bizarres and Fancies.
4. **Malmasion –** Plants are stiff with massive habit. The buds are rounded and flowers are large, generally in pink colour with good fragrance.

Origin and Distribution

The carnation is originated from Mediterranean region. During 16th century its improvement work was started in various countries. The generic name Dianthus comes from writing of Theophrastus who lived about 300 B.C. Carnation distribution occurs in Colombia Australia, *etc.*

Varieties

1. **Standard**- Lipstick, Design, Impair, Romana, Dark Tempo, White Kundra, Monaco, Regina, Cobra, Papaya, Monopole, White Shim, Nora, Red Diamond, Peppermint, Pink Eye.
2. **Spray Variety**- Ariosto, Flair, Famba, Stylo, William Shim, (M.H. Pune), Candy, Espana, Light Pink Candy, Murga, Scania, White Candy, (Solan HP), Sam-Pride, Scarlet Elegance, Arthur-Shim, Dusty and Scania (Kalimpong, West Bengal).
3. **Others** - Barbara, Silvery Pink, Natilu, White Barbara and Star Dust.

Varieties on the Basis Colour

1. **Standard Variety**
   a) **Red colour** - Dakar, Impala, Master, Nelson.
   b) **Pink colour** - Barlingna, Petra.
   c) **Yellow colour** - Pinto, Tahiti, Esty.
   d) **White colour**- Sonsara, Bogota, Riversa.
   e) **Orange colour** - Riversa, Amstel, Opinto, Solas.
2. **Spray Varieties**
   a) **Red colour** - Darling, Alistair, Ronny.
   b) **Pink colour** - Happiness, Rossini, Fantasia.
   c) **Yellow colour** - Castillo, Coreno.
   d) **White colour** - Close-up, West Christly.
   e) **Orange colour** - Nicky, Sintonia, Target.

Variegated Variety

1. **Standard Variety**
   a) **White colour** - Aledo, Forever, Olympus.
   b) **Yellow colour** - Superstar.
   c) **Orange colour** - Orange Prestige
   d) **Red colour** - Nicole.
2. **Spray Varieties**
   a) **White colour** - Regis, Saffora.
   b) **Yellow colour** - Goss Field, Naomi
   c) **Orange colour** – Sintonia
   d) **Red colour** - Challenger.

Varieties Suitable for Pot Culture

Cerotop, Charmtop (red), Maldeves (pink), White Sunny (white).

Variety Developed from IIHR

Arka Tejas, (*Dianthus caryophyllus x Dianthus chinensis*). Arka Flame (Mutant 40 Gy gamma ray).

Soil and Climate

Sandy loam soil rich in organic matter, contained with a pH of 6 – 6.5 are most ideal. Root system of carnation is very shallow and fibrous and hence it prefers porous, loose and friable soil.

Generally carnation is grown as the protected condition for commercial production. The design and orientation of greenhouse is great and important. The greenhouse should have the ridge true north and south. Plants grown in running bed in the same direction with bed 1 - 1.2 m wide and path 60 cm towards the side wall. Polyhouse or greenhouse fitted with fan and pad system can bring down temperature by 8-10°C. The portable tunnel size (3 × 1.5 × 1.5 m^3) has been found to be useful for protecting the crop. Carnation is a long-day plant, early flowering can be taken during long day than short day. Optimum night temperature during winter season should be 10-11°C, spring season 12.7°C and summer season 13- 15.4°C. Reduction in temperature 32 - 15.4°C increases the colour intensity of the flower. The greenhouse CO_2 level is 300-500 ppm on cloudy day and 800-1000 ppm on summer day. It requires 21.5 lux light intensity.

Propagation

Annual carnation is propagated through seeds whereas perennial carnation is propagated through terminal cutting. Generally carnation is multiplied with terminal cutting. Cutting of mother plant can be established in cool month of Oct - Feb. The cutting 10-15 cm long having 4 - 5 pair of leaf are broken from mother plant by hand and planted in sand bed or coco pit mist chamber. The best time for planting of cutting is Aug - Sep to get production of flower in winter and spring season.

Land Preparation

Commercial production of carnation is done in the container system of planting, for this preparation of bed is done in the size of 1m length and 30-40cm wide and 30cm above the ground level. The mixture of three part sand and one part FYM should be mixed thoroughly in the bed. Soil should be sterilized with the help of formaldehyde and solarize by using black polythene. The rooted cuttings are planted in the sterilized bed. The planting of carnation cutting at appropriate depth, 15x15 cm distance should be maintained for closer spacing and 20x15 cm for taller plant.

Manures and Fertilizers

Balanced and regular supply of the nutrient to carnation plant helps in better growth and flower production. NPK requirement of carnation plant throughout the cycle is about 20:20:10 gm/m^2 area with 8-10 kg FYM applied at third week after planting. Carnations are very sensitive to boron deficiency. It may cause excessive calyx splitting and abnormal opening of flower bud.

Planting

The carnation performs well in planting the rooted cutting which are planted on raised bed with spacing of 20 x 20 cm for standard carnation and spacing of 30 x 30 cm is ideal for spray carnation. In high density 15 × 15 cm spacing is adopted under protected condition (50plants/m^2). The ideal time for establishing mother plant is from Oct-Feb.

Irrigation

Carnation needs regular supply of moisture due to shallow root system. First light irrigation should be done just after transplantation. Drip irrigation is preferred for regular irrigation because sprinklers are not used due to encouragement of the fungal diseases. After establishment of the plant regular irrigation is essential at 5-7 days interval.

Cultural Operation

1. **Pinching-** There are two type of pinching in carnation for maximum production of flowers, *i.e.,* single pinching and double pinching. Single pinching should be done once at 5 nodes stages whereas, the double pinching should be done followed by single pinching when all shoots are 6-8 cm in length.
2. **Flower regulation-** The main objective of flower regulation is to produce excellent quality of flower in huge quantity at the time when demand is high.
3. **Staking-** Carnation is a herbaceous plant and it requires support to prevent the bending of the plant. The support is given by bamboo stick.
4. **De-shooting-** The undesired shoot of the flowering stem are removed with hand when plants are 4-5cm in height.
5. **Disbudding-** For taking the large size flower disbudding operation should be done. The side bud to should be removed just after appearance without damaging the leaves and the stem.
6. **Calyx banding-** Calyx splinting is serious problem affecting quality of flower in carnation. This can be reduced by placing a band around calyx of flower bud when they have just started flowering.

Harvesting and Yield

Carnation plants produce cut bloom offer 16-20 weeks and it varies according to pinching method followed. The optimum stage of standard carnation for distant market is brush stage when petals have started to elongate out side the calyx. The spray variety are harvested with two open flowers on each stem. Harvesting should be done in morning. Flowers are harvested with a sharp knife. The cut flower should be immediately placed in Sodium hypochloride solution @ 1ml/10litre of water.

The yield of carnation mainly depend on variety, pinching and also on the period of crop is harvested. Spray type carnation produces more flower than standard type. Generally 200 flower/m^2 can be obtained from standard type while 250 flower/m^2 can be taken from spray type.

Insect Pests

1. **Aphid** (*Myzus persicae*)- They suck the sap from the leaves and sticky substances are deposited on the leaf surface and shooty mould develop. It

can be successfully controlled by application of Rogar 1ml/litre of water.

2. **Thrips** (*Thrips tobaci*)- Thrips are the sucking type pest. They suck the sap causing leaves turn yellow. It can be controlled by application of Monocrotophas 1ml/litre of water.
3. **Red spider mite** (*Tetranychus uriticae*)- The mites generally appear in the lower surface of the leaves. They suck the sap and leaves become yellow and have a dusty coating. Its attack is very common during the summer season. It can be controlled by Dimethoate 2ml/litre of water.

Fungal Diseases

1. **Wilt** (*Fusarium oxysporum*)- It is very devastating disease and appears during hot weather associated with high humidity. The fungus is soil-borne therefore sterilization of soil is very essential. It can be controlled by application of Bavistin 0.2 per cent or drenching of soil with Dithane-M-45 0.1 per cent.
2. **Alternaria leaf spot** (*Alternaria dianthii*)- It is very common foliage disease. The fungus causase black spot on the leaves and stem and in case of severe infestation plants die. It can be controlled by spraying of Capton 0.2 per cent or Indofil-M-45 0.2- 0.3 per cent.
3. **Gray mould** (*Botrytis cinerea*)- This is the most common disease in carnation. The high humidity is most favourable for fungus growth and it can be minimized by lowering the humidity.

Physiological Disorders

1. **Calyx splitting**- As a flower bud open and petals reach to their full size the calyx may split down either half or completely. The temperature different of 9°C between day and night (10-18°C) is also one of the important factor of splitting. Low nitrate, high ammoniacal nitrogen during low light period or low Boron level also enhances calyx splitting. Varieties like Espana, Carburet are less prone to this problem.
2. **Slabside**- This disorder increases in cooler period and bud does not open, so petals grow on the one side only. It is controlled by the gradually increasing of optimum temperature.

Chapter 33

Orchids

Introduction

Orchids are the most beautiful flower and have conquered the cut flower industry all over the world. They are valued for cut flower production and as potted plant. Orchids are the derived from the Greek world *Orchis* means Testicles. The word *aorchis* was first used by Theophrastus (372-287) in his book "De Historia Plantarum" (The natural history of plants). He is considered the father of botany and ecology. It belongs to important monocotyledons series microsperma. It belongs to Orchidaceae family which includes about 800 genera and 35000 species and is the largest family among flowering plants. Some important genera of orchids which are basic and 2n number of chromosomes are Arachis (X=19, 2n=38), Cattleya (X=20, 2n =40), Cymbidium (X=40, 2n=80), Dendrobium (X=19, 2n=38) and Vanda (X = 19, 2n=38). Orchids are evergreen in temperate regions however main region are found Indo-Malayan and Tropical American region like Brazil and Mexico. In India main region are Himalaya region like Sikkim, Meghalaya, Tripura, Assam and Bengal. Orchids are perennial, terrestrial, epiphytic, saprophytic and intermediate herbs with rhizomes or pseudo bulb or tuberous root or aerial epiphytic.

Botanical Description

According to the growth pattern orchids can be divided into two basic growth type.

1. **Monopodium (Monopodial) one footed-** They have single non-branching stem growing year after year and bears flower on lateral branches. They have vertical growth habit, absorbs nutrients and moisture from air and leaf, *e.g., Renanthera, Vanda, Phalaenopsis and Arachnis.*
2. **Sympodium (Sympodial) many footed-** The sympodial orchids have a horizontal growth habit and the new growth develop from rhizome. In which main axis comprises of annual portion for successive axis each of which bears scale leaf and terminal flower. Some genera belonging to this group have thickened stem called as pseudo bulb, *e.g., Cattleya, Dendrobium, Cymbidium.*

Orchids can be divided into 4 types according to flowering condition.

1. **Epiphytes-** These orchids grow on the tree and fasten their support with help of clinging roots which arise from rhizome and from network between orchids and support and form a reserve for humus, *e.g., Vanda, Dendrobium.*
2. **Lithophytes-** These grow on moist and shaded rocks and crevices of stony walls, *e.g., Diplomeris hirsute* and *Hernanium josephii.*
3. **Saprophytes-** These orchids lack green leaf and process and fleshy underground rhizomes with or without roots. The rhizomes are usually branched and absorb moisture from humus soil. Endotrophic mycorrhiza occurs in almost all saprophic orchids.

4. **Terrestrial-** These are sympodium and usually have a rhizomes or tuberous root or pseudo-bulb. These are xerophytes or mesophytes which are perennial in nature, *e.g., Cymbidium, Liparis.*

(A) Monopodials

Arachis genera	- Catherine, and Ishbel.
Renanthera genera	- Kilauea, Poipu and Tom Thumb.
Phalaenopsis genera	- Peppermint, Canary, Violet Mist, Zada, Tammy and Texas Star.
Vanda genera	- John Clubb, Bill Sutton, Ellen Noa, Evening Glow Honolulu, and Onomea.

(B) Sympodials

Cattleya genera	- Bow Bells, Empress Bells, Estelle and White Christmas.
Dendrobium	- Somia-17, Sonia 28, Kasem Gold, Kasem White, Emma White and Snow White.
Cymbidium	- Peter Pan Greensleaves, Promona, Showgirl Cooks Bridge, Dingwall Heaves.

Soil and Climate

For terrestrial orchids a potting mixture of loam soil, reverse, leaf mould, charcoal dust, old mortar in the ratio 1:1:1:0.5:0.5 improved growth and flowering the common component of media used for growing epiphytic orchids are tree firm fiber, bark, coconut husk, brick and charcoals.

Most of the cultivated orchids thrive in a day temperature varying from 15.5 to 21°C and night temperature are 10-15.5°C based on preferable temperature range orchids may classified as warm (tropical) intermediate (subtropical) and cool (temperate) species, orchids in generally prefer high humidity. Monopodials require high humidity (up to 70 per cent) while sympodial require comparatively less humidity (40-50 per cent) and pH range are found 5.5–7.0.

Propagation of Monopodial Orchids

1. **Stem cutting-** About 40-50 cm long top cutting with at least two well developed aerial roots are ideal and intermediate cutting can also be used. If smaller cutting 2-3 nodes are used the time taken to flowering is longer.
2. **Layering-** Air layering found successful in Vanda. A slanting cut is given on the stem at about 20-30 cm below the apex. The wound may be covered with the some moist media as in the case of air layering.

Propagation of Sympodial Orchids

1. **Division-** This method involves division of large clumps into smaller unit and is suitable for Dendrobium, Symbidium, *etc.* care should be taken to see that each unit has at least 4-5 shoots including the old ones.

2. **Back bulbs-** The older shoot of sympodial orchids which are lesser active physiologically are called back bulbs. These back bulbs is cut off from the mother plant and kept horizontally over a moist medium. After some times the stick roots and sprout from the nodule regions.

Tissue Culture

a. Various plant parts like shoot tip or meristem, leaf and leaf segments, stem segments, floral parts and areal root, *etc.* have been used for tissue culture of orchids. Meristem culture is most popular and extensively used method for the commercial propagation of orchids.

b. In tissue culture when meristem are used the ex-plant are sterilized with the suitable steriland, washed in sterilized water and transferred into culture medium.

Manures and Fertilizers

The terrestrial orchid recieve nutrient from the decaying leaves and other vegetative matter. Epiphytic orchid on the other hand receive nutrient from the first rain and also from the bud dropping. In general the N:P:K is equal proportion is ideal. Ammonium nitrate, orthophosphoric acid and potassium nitrate can be use for the supply of NPK respectively. The organic manures like cow dung, neem oil cack, poultry manure, *etc.* are also used for orchids.

Planting Care

Planting Monopodial Orchids in Container

1. Monopodial orchids have a vertical growth and in general they are not planted in pots.
2. After filling the pots with the media the plant is the placed in the center of pots and support with a stick. The pot can be kept on the ground or can be hung. Dwarf vanda are suitable for growing in containers.

Planting Sympodial Orchids in Container

1. The sympodial orchids are planted in such way that the oldest shoot touch a side of containers and the growing point faces then center of the containers. The sympodial orchids are also used for growing in pot or hanging pot or baskets.
2. Usually they are planted in two rows at spacing of 20-30 cm between plants and row. While planting in trench individual stacking is not necessary.

Irrigation

The moisture requirement is different in all the genera of the orchid. Monopodials require less water, while terrestrial types need abundance of moisture but with good drainage and aeration. The maintenance of adequate humidity in the atmosphere and moisture at the root level is the most important aspect of orchid cultivation. Plant are mainly lost not because of inadequacy of water but because of their overdose.

Cultural Operation

1. **Replanting/repotting-** When the plant has filled the pot and there is no room for further growth and in case of field planting of sympodial, over crowding condition is necessitate replanting.
2. **Regulation of light/shade-** Providing optimum shade is an important aspect of cultivation of orchid. The vase life of orchid increases in 25-50 per cent in double level of shade.

Harvesting and Yield

Flowers are harvested twice in a week during peak production period and once a week during low period. Harvesting is done when 30-40 per cent of flowers are open and harvesting should be preferably done in evening.

Orchids have a long gestation period and it produce flower after 3-4 years of planting, *i.e.*, Dendrobium 1-2spike/plants, Vanda- 4 sprays/plant with 10-15 flower/sprays and Symbidium-2500 spike/plant from 4th year onward.

Insect Pests

Aphid, Orchids weevil, Orchids bulb borer, Orchids fly and motes are the important insect and it can be controlled by Malathean, Althene, Methenychlore, Calthene 0.2 per cent.

Fungal Diseases

Leaf spot, phythium black rot and flower blight are the important fungal diseases and it can be controlled by blitox 0.3 per cent or bavistine 0.1 per cent dithane-M-45 – 0.2 per cent at regular interval.

Bacterial Disease

Bacterial soft rot is the important disease and it can be controlled by Streptocycline 0.01 per cent.

Viral Disease

Mosaic, Blossom brown necrotic stick are the important viral diseases and it can be controlled by controlling the main transmitting agent green peach aphid (*Myzus persicae*) by application of systemic insecticide.

Chapter 34

Gerbera

- ✰ Common name - Gerbera, Transvaal daisy
- ✰ Botanical name - *Gerbera jamesonii*
- ✰ Family - Asteraceae/Compositae
- ✰ Chromosome number - 2n = 50 (X=25)
- ✰ Origin - South Africa
- ✰ Gerbera comprises about 70 species, which are of African and Asiatic origin
- ✰ The genus Gerbera is named in honour of German naturalist Trangott Gerber.

Introduction

Gerbera is an important commercial cut flower crop grown throughout the world in a wide range of climatic conditions. Gerbera is a herbaceous flower crop with leafless stalk and daisy like flowers. It is ideal for flower beds, borders pots and rock gardens. The flowers come in a wide range of colours and land themselves beautifully to different floral arrangement. The cut blooms when placed in water remain fresh for a reasonable amount of time.

In California and Florida, it is grown in outdoors and is becoming increasingly popular as potted plant and bedding plant, whereas in other parts of United States it is cultivated in greenhouses. Gerbera cultivation has emerged as a very important option to progressive farmers in many parts of India, especially in Maharashtra, Karnataka, North Eastern states, Uttar Pradesh and Uttarakhand.

Botanical Description

Plants are stemless and tender perennial herbs, leaves radical, petioled, lanceolate, deeply lobed, sometimes leathery, narrower at the base and wider at toe and are arranged in a rosette at the base. Flowers are solitary daisy like grown in a wide range of colours. The height of plant may be up-to 45 cm and may spread up-to 60 cm.

Flower heads are solitary, composite (many flowered) having conspicuous ray floret in 1 or 2 or more rows. Inner row is very short and tubular and 2 lipped. Flowers are like daisy in many clear or bicolours, *e.g.,* White, yellow, pink, orange, brick red, scarlet, salmon, terracotta, *etc.*

Origin and Distribution

Gerbera flower is native to South Africa (Natal *G. jamesonii* and cape- *G. virdifolia* were crossed with each other to produce modern types) and Asia. At the turn of the century, gerberas were cultivated in England, Belgium, USA, Germany, Australia, Italy and India. In India gerbera is widely found in temperate Hamalaya's from Kashmir to Nepal at the altitudes of 1300 to 3200 mts.

Varieties

Gerbera is very rich in its varietal wealth due to ease in hybridization and survival of seedlings. Thousand of cultivars have been developed which are highly suitable for commercial production of cut flowers under open as well as under greenhouse conditions and several varieties suitable for pot culture.

According to flower colour also cultivars are as follows:

1. **Orange**: Alegro, Dune, Goliath, Mistique. Orange Glame, Sunway
2. **Pink:** Calcutta Pink, Casva, Citella, Delama, Devas Memory, Dokevita, Evening Bell, General Kaisar, Glory, Kalimpong Pink, Mella, Pink Elegance, Pink Star, Rosalin.
3. **Red**: Calcutta Red, Camilla, Cango, Diablow Gateway, Kalimpong Red, Ruby Red, Salvadore, Sangria, Zingaro.
4. **White:** Balance, Calcutta White, Dalma, Glaria, Pride of Sikkim, Twiggy, Winter Queen.
5. **Yellow**: Arka Krishika, Alesmara, Aruba, Black Heart, Calcutta Yellow, Carrona, Dana Ellen, Goddongak, Golden Gate, Indus Kumari, Kalimpong Yellow, Paradiso.

According to flower size also cultivars are grouped into tall, medium and dwarf types are as follows:

1. **Tall variaties**- When flower stem length is more than 50 cm, *e.g.,* Funda. Diablo, Sunanda, Dark Red;
2. **Medium varieties-** When flower stem length is between 40-50 cm, *e.g.,* Ornella, Sun Set, Thalassa, Tavila, Kolika, Ruby Red, Pink Elegance, Rose Bella,
3. **Dwarf varieties-** When flower stem length is between 30-40 cm, *e.g.,* J.S. Lal, Alemasa. Y. Record, Janson Hybrid, Verasace, Indu Kumari, Yellow Queen, Gen Khaiser, Nebulusa, Red Monarch, Sangria.

Under Indian conditions, the popular gerbera varieties amongst commercial growers are- Jaffa, Sangria, Rosula, Oprat, Romona, Salina, Tecora, Starlight. Aida, Ventur, Ornella, Aalasmeor, Goliath, Pink Elegance, Rosalina, *etc.*

Soil and Climate

For growing gerbera successfully, soil should be adequately porous and well drained. Thus light soils with good organics contents are more suitable than heavy soils. The ideal soil pH for gerbera growing is 6.0 to 6.5 but it can tolerate little higher pH, *i.e.,* upto 7.5. Higher pH of soil needs its correction to the desirable limits. Before starting a cultivation of gerbera, soil sterilization is absolutely necessary especially against fungus like phytophothera, cryptogera and nematodes which cause serious wilt disease and may destroy a crop partly or even completely. For this formal dehydre or methyl bromide gas can be used and soil is covered with polythene for 48 hrs. Therefore, as soon as selecting the site, get the soil analyzed to decide its further reclamation.

Gerbera needs mild climate for its vegetative growth and flowering. Extreme of cold or hot temperatures are not desirable. Therefore, under tropical and sub-tropical climate, gerbera is grown in open or in unheated plastic houses. In temperate climate, gerbera plants need protection from frost and are grown in greenhouses. Gerbera plants require sunny situation in mild weather for better plant growth and flowering but during summer months they should be lightly shaded, if grown under open conditions. Apart from significant improvement in quality, flower yield increases by 50-80 per cent under protected conditions compared to open cultivation. To control light intensity and solar radiation, white shade net (50 per cent) is used. Approximately 35,000 to 40,000 lux light intensity is required on the plant level.

It is observed that gerbera produced more flowers/m^2 in a cooler greenhouse when temperature maintained at 18°C day and 12°C night than 20°C day and 16°C night temperatures. During entire growing and flowering period high humidity should be maintained. If the air humidity drops below 50 per cent, the young plants suffer. The humidity can be maintained more easily in closed structures

Land Preparation

In general, gerbera are grown on raised beds to assist in easier movement and better drainage. The land should be deeply ploughed 2 to 3 times and brought to a fine tilth. The beds should be raised at 40 cm height and the width of beds should be maintained at 1 to 1.2 m leaving 30 cm between two beds. Adding of two part well decomposed farmyard manure, one part of sand and one part of coconut coir pith or paddy husk. At the time of bed preparation (after fumigation) neemcake (@ 1kg/m^2) is added as prevention against nematode. All material should be mixed thoroughly for optimum result.

Propagation

Both sexual and asexual methods are employed in propagation of gerbera. Propagation through seed is used to breed new varieties by breeders. Gerbera seedlings exhibit variation when obtained from open pollinated seeds or in a planned programme useful variations are released as new varieties. It is a long processes and it takes 3-4 years to bloom a seedlings. Care should be taken that seeds should be sown immediately after harvest otherwise they loose viability quickly.

Vegetative methods like division of clump, cutting and micro-propagation are very commonly used.

1. **Division**: It is most common and widely adopted method of propagation. In this method the large clumps are divided into seedling plants and are set in the field. Before planting, roots and leaves are trimmed keeping

central axis intact. While setting the suckers care should be taken that soil not cover up the central axis. This method is generally practiced in June, when the plants may be set out in field. Before transplanting, the roots and leaves of suckers are trimmed, keeping the central shoot intact. Care should be taken that the soil does not cover up the central growing point while setting the suckers in new bed.

2. **Cutting**: Gerbera is also propagated through cuttings. Healthy plants are kept without water for 3-4 weeks. Then roots are pruned and planted in peat and held at 25°C with 80 per cent relative humidity. The buds in the axils of leaves are detached and rooted in rooting medium. Plants become ready for transplanting in 2-3 months. Approximately 40-50 plants can be produced in 2-3 months from a single mother plant. Young stem cutting produce roots and shoots much easily and quickly under misty condition.
3. **Micro propagation**: Normally, gerbera is propagated through seeds and division, but with the availability of modern techniques of tissue culture of plants which allows the production of disease free planting material in unlimited number from single explants. It offers a very good opportunity for large-scale multiplication of gerbera. It involves rapid multiplication of explants by repeated sub-culturing and preparation of divisions for transfer to soil. Shoot tips, inflorescence buds, flower buds, flower heads, midribs, *etc.* have been employed as explants for micro-propagation. Various culture media for explants from different sources have been found useful in obtaining the maximum number of plantlets.

Manures and Fertilizers

The basal nutrition applied in gerbera are on the basis of soil condition. Gerbera prefers organically rich medium for proper growth and flowering. Therefore, addition of organic matter in medium results in healthy plants and excellent flower production of better quality of flowers. Phosphorus and calcium are best given before planting as a basal dressing. Depending upon the Ca levels in the soil, P can be given as TSP (triple super phosphate) or SSP (single super phosphate). Soil analysis will decide the quantity of organic matter and other fertilizers to be added. Generally in fairly sandy soil, addition of 7.5 kg/m^2 well rotten farm yard manure gives desirable results. Gerbera responds well to the applications of NPK which improves the numbers of leaves, suckers and large number of blooms of big size. Excess of N reduces of flower quality and flower production. Generally, N@15 gm was found to be optimum for better flower production of quality bloom. N, P and K in the ratio of 3:1:1 produced highest flower yield.

Magnesium can be given in the form of finely powdered dolomite limestone if pH also needs to be raised. Magnesium can also be added in the form of water soluble magnesium sulphate either before or after planting.

Till 2-3 weeks after planting, no fertilizers are applied. After 3 weeks of plantation application of N:P:K (19:19:19) at 0.4 g/plant every alternate day with EC 1.5 mS/cm to be done for first three months during the vegetative phase to have better foliage. When flowering commence application N:P:K (15:8:35) at 0.4 g/plant every alternate day with EC 1.5 mS/cm should be done for more flowers and better flower quality.

Planting

In greenhouses or in open field conditions, two rows or 4 rows planting system are generally followed. Hence, accordingly bed width is kept. Gerbera can be planted all the year round but there are two common planting time, first in spring, *i.e.*, January-February and second one is summer, *i.e.*, June-July. It is observed that planting of gerbera from second week of May to second week of July produced more number of cut flowers. For large flowering gerberas 8-10 plants/m^2 are planted. Close planting is followed for small flower cultivars. The spacing in rows ranges from 30-40 cm and plants are planted at 30-37.5 cm apart was found to be optimum for better leaf growth, flower stalk length and flower diameter. While planting, care should be taken that crowns should be well above the soil surface. Planting time varies with place of growing, its climatic conditions and method of growing.

Irrigation

Gerbera is a moisture loving plant and hence, it needs thorough irrigation at regular interval. However, wet feet are not desirable so avoid waterlogging conditions. Water requirement vary with the varying environmental conditions. Gerbera, a deep rooted crop requires thorough irrigation immediately after plantation. Overhead irrigation to be given to plants for three weeks to enable uniform root development. Irrigation methods like flood irrigation, sprinkler and drip irrigation, can be employed according to the facilities and finances available. The water requirement of gerbera plant is approximately 300-700 ml/plant/day depending upon the season.

Cultural Operations

1. **Weeding and hoeing:** Gerbera being perennial crop and remains in the field for 2-3 years and its rosette nature of growth, weed appears only in the early stages of crop. Chemical like Trifularin, EPTC, nitrolen and diphenamid after one day of transplanting markedly reduced the growth and population of weeds. Two to three manual weeding are required at 15 days interval. Efforts should be done to keep the soil loose and friable by regular hoeing and free from weeds.
2. **Flower regulation:** After planting, normally it takes 2-3 months to produce flower by the plants. If plants take longer period for flower production. Plants can be sprayed with 100 ppm of $GA_{3.}$ This will result in early flowering with longer stems.

Harvesting and Packing of Flower

The harvesting of gerbera flowers should be done at optimum stage. Gerbera is a 24-30 months crop. The first flowers are produced 7-8 weeks after plantation when plants are with 14-16 leaves. In general, flowers are harvested when fully open and disc flowers in the two outer rows shed pollen. They respond very well to re-cutting of stem before placing in water or preservative solution. Flowers should be pulled and placed immediately in water. All the flowers are packed in insulated boxes individually for safe transport. Good quality gerbera should have at least 40 cm stem length and should be firm and straight. Stems showing bending are not desirable in the trade. The flower should be uniform in size and should not be less than 7 cm in diameter. Flowers of gerbera are packed in flat boxes having paper shreds and holes for individual flowers.

Yield

The yield of gerbera flowers is greatly influenced by climatic conditions or environment of poly-houses, cultivars and management practices followed during cultivation. The yield under greenhouse is around 200-250 flowers/m^2/year (6 plants/m^2) of which 85 per cent are of first grade quality.

Post-Harvest Management

The stems should be put in clean buckets with clean water immediately after harvesting and place them in a cool area. Before every use these buckets need to be disinfected to avoid the growth of bacteria in it. Bacteria block the stem so that it cannot take up any water. Use of clean water is important. Various flower preservatives are found beneficial to enhance the vase life of gerbera flower. The recommended floral preservative on gerbera are $AgNO_3$ (20-30 mg/l) +sucrose (3-6 per cent), $AgNO_3$ (20 mg/l) + sucrose (2 per cent) + $NiCl_2$ (150 mg/l), HQC (200 mg/l), + sucrose (3 per cent) and DICA (50 mg/l) + sucrose (2 per cent).

Main problems observed in field or greenhouse:

1. If flower stems are too tall or foliage is too large, increase light intensity or use B-9.
2. If flowers are too short or hidden in foliage, this condition could result from excessive fertilizer (nitrogen or ammonia).
3. If plants are dying out too frequently, too much growth regulator or average growing temperature is too low.
4. If flowers are distorted, it is caused by mites or thrips.
5. If plants fail to grow, usually caused by poor drainage, packing soil too tightly or low soil temperature.
6. If plant wilts or dies, it might be planted too deep and eventually was killed by crown rot.

Insect Pests

1. **Whitefly** (*Trialeurodes vaporariorum*)- This is a sucking type pest. Its infestation is more in greenhouses. In severe cases yellowing of leaf and it feeds on the lower side of leaves excrete large quantity of honey dew which leads to development of black sooty moulds on the leaves. It can be controlled by application of rogor (dimethoate) at 2 ml/l.
2. **Leaf minor** (*Liromyza trifolii)*- It is a serious pest of gerbera. The adults lay eggs on the leaves and larvae bore into the leaf and make irregularly shaped tunnels or blotches, which are generally light yellowish tan to brown in colour. It can be controlled by application of Dimethoate 0.1 per cent.
3. **Mites** (*Hemitarsonemus latus* and *Steneniotarsonemus pallidus*)- This is the serious pest of gerbera. The infectation is more in greenhouse conditions. Development of leaves and flower buds are adversely affected and malformed flowers are produced. It can be controlled by sprays of kelthane (1 ml/l) and thiodas (2 ml/l).
4. **Aphids**- This is the sucking type pest and suck the sap from the young leaves and newly growing buds which results in distortion of tissue and finally yellowing of the leaves. It can be controlled by sprays or fumigation by dichlorvos 0.1 per cent or oxamyl 0.05 per cent.

Diseases

1. **Crown rot** (*Phytophthora cryptogea*)- This disease is caused by soil born fungus and in severe cases plant show the wilting like symptoms. It can be controlled by drenching of soil with bavisteen @ 2 g/l of water, regular crop rotation and spray of Blitox (Copper oxychloride) at 1-1.5 g/l.
2. **Root rot** (*Pythium spp.)*- It is a soilborne disease caused by *Pythium* and *Rhzoctonia spp.* In severe cases the plant become stunted in growth, dropping of younger leaves and ultimately drying of the entire plant. It can be controlled by the spray of carbendazim 2 g/l or benlate (benomyl) 3g/l.
3. **Anthracnose** (*Colletotrichum gloeosporoides*)- It is a very serious disease in gerbera crops. Some time it is the limiting factor for the production. The important symptoms are circular, scattered, reddish brown spots and resulting in withering, rolling and drying of leaves. It can be controlled by spraying with carbendazim (0.1 per cent).
4. **Blossom blight** (*Botrytis cinerea*)- Cool moist weather of prolonged rains is the most favourable condition for disease encouragement. Light brown irregular water soaked up area appears on flower stalks, which enlarge and coalesce producing distinct depressed lesions. It can be controlled by benlate @ 0.1 per cent or thiram 0.2 per cent.

5. **Powdery mildew** (*Erysiphae cichoracearum* and *Oidium ersiphoides*)- The effected parts are covered with white mycelial growth. Due to white growth photosynthetic area is reduced. The diseased portions get dried and fall. It can be controlled by application of carbendazim (0.1 per cent), sulfex (0.3 per cent), karathane (0.5 per cent).
6. **Bacterial blight**: The disease is characterized by small to large circular or irregular, brownish black leaf spots with or without concentric rings. It can be controlled by streptocycline (0.2 ml/l).

Physiological Disorders

1. **Pre-harvest stem break**: Pre-harvest stem break occurs when plants are allowed to wilt during the day or when day temperature increases rapidly during bright sunny weather. Under such conditions, the flower stem is subjected to stress and wilts. After watering of cooler growing conditions, the water to the stem is rapidly replenished and the stem becomes turgid again. During rehydration, extreme conditions are imposed on cells in the stem where rapid elongation is occurring. This part of the stem accumulates the highest water content. Keeping soil moist during the heat of the day or reducing air temperatures can minimize pre-harvest stem break.
2. **Premature flower wilt**: Premature wilting of the flower occurs while stems are still attached to the plants and often develops just as petals are in full expansion. The cause of problem is suspected to be a lack of storage carbohydrates needed to attain the integrity of the rapidly developing flower. It occurs mostly after the period of cloudy days with low intensities followed by a clear sunny day. If possible screening of the varieties should be done, planting of cultivars with lesser effective to this disorder is advisable.

Chapter 35

Lily (Lilium)

- ☆ Common name - Lilium, Easter lily
- ☆ Botanical name - *Lilium longiflorum*
  Lilium lancifolium, L. auratum, L. cendidum, L. elegans, L. formosum, L. nepalense, L. philippinense
- ☆ Family - Liliaceae
- ☆ Chromosome number - 2n = 24, 36 (X=12)
- ☆ Origin - Northern hemisphere
- ☆ Lilium is a bulbous plant
- ☆ Lily is the national flower of Italy

Introduction

Lily is being cultivated for centuries as an important ornamental bulbous flower. It is commercially cultivated for cut flower production. The flower stems of lily are long and it has longer vase life. Lilies are very useful flowering plants which can be planted in any location as they do well in the bed or border with other perennial which can be found in temperate regions of the northern hemisphere.

Lilies are wonderful ornamental plants with varied uses, grown in border, beds, pots and are excellent cut flowers of magnificent appearance and beautiful colours. Asiatic and Oriental lilies are leading cut flower group and in the international market it is dominated by the Netherlands. Oriental and Asiatic hybrid lilies are compact, with plant height of 60 to 150 cm that bears many flowers.

Botanical Description

Lily belongs to genus *Lilium* and it has about 400 species. They have broader leaves than most monocots, many with the networking veins usually associated with dicots. It is herbaceous, perennial, bulbiferous plant. Leaves alternate, rarely whorled, sessile or subsessile, usually linear to linear-lanceolate. Lily have some of the showiest flowers in the plant world. Flowers are terminal solitary, recemose or umbellate and perfect. Their large colourful tepals come in white, yellows, brown, oranges and reds, blue and violets. Tepals 6, free, usually convenient, sometimes strongly recurved or revolute, white, yellow, greenish or reddish to purplish, nectariferous near base adaxially.

Origin and Distribution

Lilies are natives of the northern hemisphere up to South Canada and Siberia and their southern limit is Florida and the Nilgiri mountains of India. It is distributed to east coast of Asia, the west coast of North America and the Mediterranean region. Species like *Lilium nepalensis, L. wallichianum and L. polyphyllum* are native of Himalayan region and *L. niligiriensis* have originated from Nilgiri hills. A few species have been found in North-East India also.

The first authentic record comes from Assyrian monuments dating back to 1000 BC with sculptured lily forms. The Palestinian coin makers were stamping lily on shekels before 143 BC. *Lilium longiflorum* (Easter lily) became a florist flower which existed in the Japanese gardens before trade opened the orient to the west. After that it was brought to the England by the Royal Horticultural Society in 1819. A number of lily bulbs were imported by the US from the Netherlands for both professional forcing and garden usage. In India, lilies are mostly grown in the hills but *L. longiflorum* do well in the plains, especially under subtropical regions.

Varieties

Asiatic Cultivars

1. **White:** Alaska, Lucyda, Marbelle, Pulsar, Sancerre, Ventoux.
2. **Red**: Avignon, Grand Paradiso, Monte, Negro, Nerone.
3. **Pink:** Azurra, Chianti, Geneve, Rembrandt, Renee, Sorbet, Tiareno, Toscana.
4. **Orange** Apledoorn, Elite, Loreto, Menton, Prato.
5. **Yellow**: Adelina, Grand Cru, London, Mona, Nove Cento, Parma, Polyanna

Oriental Cultivars

1. **White**: Casa Blanca, Dream, Mont Blanc, Montreal, Primeur, White Sheen.
2. **Red**: Corina, Jazz, Red Carpet, Strangazer.
3. **Pink**: Tiara, Acapulco, Alena, Berlin, Le Reve, Marco Polo, Nigata, Pesaro.
4. **Yellow**: Butter Pixie, Chicogo, Dream Land, Yellow Blaze.

Soil and Climate

Lily cultivation prefers a well drains and sandy loam which pH ranges between 6.0-6.5. Heavy loam and clay loam are not suitable but can be used if humus retaining substrates are added to a depth of 30 cm in the soil as this will aerate the top soil. The growing medium must be porous for good aeration and water drainage. Potted lilies have been forced successfully in several types of growing media. Mixture of potting media should be sandy loam 50 per cent, sphagnum peat 25 per cent and sand 25 per cent.

Lily are sun loving climate. But root system of lilies cannot tolerate direct sunlight. Optimum day temperature for quality flowers should be 15-20°C, whereas night temperature of around 8-10°C is suitable. Soil temperature should be below 20°C. It is observed that 10 to 15°C night temperature and 18 to 22°C day temperature is optimal for Asiatic lilies. Oriental hybrids need warmer night temperature of 15 to 18°C and day temperature nor exceeding 30°C. Lilies require some shading.

Propagation

Generally lilies can be propagated by seed and asexually through vegetative propagation (from bulblets obtained from bulb scales, by bulb division, from hypogeal bulblets, from bulblets obtained from leaves and from bulbils).

1. **Bulblets:** Lilium are generally propagated by division of bulblets. A lily bulb from suckers at the base of the stem just above the mother bulb. These bulblets can be carefully cut away with a sharp knife and planted directly in the beds. Some of the longer ones may take one year to produce flower after being separated from the mother bulb whereas smaller may take 2 years to bloom.
2. **Bulbils:** Bulbils of tiger lily do not have a dormant period when cultivated at temperature of more than 0°C but under open cultivation at 10°C they become dormant. Plant grown from bulbils produced successive adventitious roots, each with a life not exceeding 4 months. After flowering, some species of Lilium especially *L. lancifolium* produce black bulbils, they are about size of peas. They are formed in the axils of leaves. Other species like *L. wallichianum, L. sargentae* and *L. bulbiferum* also produce bulbils. These bulbils are gathered carefully in the late summer and immediately sown in the beds. Bulbils take 3-4 years to reach blooming size.
3. **Bulb/scales**: This is a most common method of multiplication especially in the varieties that do not produce bulbils. Scales are treated with fungicides and drained thoroughly and later placed in plastic crates, between alternate layers of peat or vermiculite. The crates are stacked in a dark room where the optimum temperature and humidity are maintained. Scales are treated with NAA or IBA for increasing rooting and number of bulblets.
4. **Seed**: Seed propagation is not a commercial practice for ornamental lilies. It is an easy and cheap method of raising lilies. One big disadvantage is that new plant could take one to five years to reach flowering. Seeds may be sown indoor in flat pots during the winter and planted outdoor in the spring. A good soil mixture for seed sowing consists of 7 parts loam, 3 parts peats and 2 part sand.

Land Preparation

Field should be dug out properly. Organic matter may be supplied at 4-5 kg/m^2 in the form of well decomposed farmyard manure. The beds should be fumigated or treated with aerated steam to control the insect and mite pests, disease organisms, nematodes and weeds. Good drainage is essential factor as there is loss of lily bulbs if water stagnation occurs in the field. Drainage should be provided for satisfactory growth and flowering in lilies.

Manures and Fertilizers

Application of nutrients is essential for proper growth and flowering of liliy. Nitrogen should be applied at 1 kg of calcium ammonium nitrate/100 m^2 after three weeks of planting of bulbs. If plants are weak and showing deficiency of N, then top dressing of fast release nitrogen (urea) at 1 kg/100 m^2 may be done before three weeks of flower harvesting.

Planting

The need of planting clean, healthy and disease free bulbs can hardly be overstressed. Time, depth and method of planting depend upon the type of lily to be grown, although there are certain general rules for all members of the lily groups.

Oriental hybrid lily bulbs of 16-18cm or larger are usually taken for cut flower production. Smaller bulb less than 12cm should be avoided. For Asiatic lily 20-22 cm circumference bulbs are recommended. It is better to use smaller-sized bulbs if climatic conditions are favourable. It is advisable to plant lily bulbs on raised beds 15 cm above the ground level. Bulbs are generally planted at 6-8 cm depth during winter. The distance of bulb to bulb is generally kept 15 cm, whereas 25 cm distance is kept between two rows. Irrigation channels of 45 cm in between the beds of one meter wide to convenient length should be made. Bulbs are planted during October-November in the plains and it bears flower during January-February.

Irrigation

Irrigation is the most important for the lily crop, especially during the first three weeks after planting. Watering must be carried out sparingly. Sudden moisture deficit is likely to increase the number of blasted bud. Excessive watering is harmful to crop and it checks the proper availability of the oxygen in root zone. During dry period consumption could increase 8 to 10 litre/m^2

Cultural Operations

1. **Weed control**: Shallow hoeing followed by weed cleaning is generally recommended for lily. Application of certain herbicides has been found effective to control weeds in lily field. Chloropham at 3.5 1/ha and propyzamide at 2.25 kg/ha are found beneficial to control the wild population except leguminous weeds.
2. **Staking**: Staking is an important operation for tall cultivars particularly in those have more than 80 cm heights. Lilies grow to a height of 60-180 cm. Staking or netting of lily plants is required when they become 50 cm tall. Netting is to be done at 50-60 cm intervals. It reduces breakage of stems.

Flower Forcing

Flower forcing is an important operation to induce flowering at a desired period with the goals of producing flowers during off-season and specified date and for sale at higher prices than during normal blooming season.

Vernalization accelerates growth by stimulating shoot emergence, internodes elongation and rapid leaf unfolding. About 6 weeks, storage of bulbs at 4.5°C was found ideal for early emergence and produced taller plants. Pre-cooled bulbs from July and August harvests produced plants with longer leaves and those from September and October harvests were shorter than those from non-precooled bulbs.

Bulbs treated at 4.5°C for 30 days followed by 21°C for 30 days had higher GA activity and shoots were slower to flower as compared with bulbs treated at 21°C for 30 days followed by 4.5°C for 30 days. Bulb vernalization and photoperiod treatments lead to rapid flowering through two induced mechanism.

Harvesting

1. **Flowers**: Flowers of lily are harvested when the first flower most bud shows full colour but has not yet opened. If lilies are harvested at an earlier stage, buds may take longer time to open, may not open completely or may be misshapen. The spike of flowers is generally cut 15 cm above the ground level so that the development of bulb may continue in the soil. After cutting of flower, they should be immediately placed into clean water with preservatives and brought to the cool room to remove the field heat.
2. **Bulbs**: When the aerial parts of the plant have died, bulbs should be dug out. If lifting of bulbs is delayed the bulbs may become more mature. Bulblets and the bulbs should be carefully harvested. They should be properly cleaned and treated with suitable fungicides like carbendazim to prevent rotting during storage.

Yield

The yield of flowers greatly varies according to cultivar, package of practices adopted during cultivation and climatic conditions. The average production of marketable flower stem is 1,00,000 to 1,12,500/ha whereas, bulb yield is 1,25,000 to 1,50,000ha.

Post-Harvest Management

Flowers

After harvesting, flowers are graded as per the number of flower buds per stem, length and firmness of stem. In the Netherlands they are graded by the lowest number of calyx per stem and the highest number of calyx per stem. Removal of anthers from open bloom is important operation to prevent the spoiling of flower or any surfaces on which it might fall. The foliage must be removed at 10 cm above from the bottom. The lilies are bunched in a bundle of six stems. After bunching, the spikes are placed in a suitable preservative and stored before shipping. The optimum storage temperature of cut flower is 2.0-3.0 °C for a shorter period.

Pulsing of cut lilies with 0.50-1.00 mM silver thiosulphate (STS) extends the vase life of flower. Holding solution of 200 ppm 8-hydroxyquinoline citrate (8 HQC)+3 per cent sucrose prolonged vase life of cut lilies.

Bulbs

After cleaning, bulbs are graded according to their circumference. In the United States, the commercial grades are 6-7, 7-8, 8-9 and 9-10 inch in size. The bulbs should be properly packed in perforated trays in peat moss or sawdust and placed at 2.0°C with 70 per cent h RH for minimum 2-3 months in the cold storage.

Insect Pests

1. **Aphids**: Five species of aphids infest this host: crescent-marked lily, green peach, foxglove, melon and purple-spotted lily. It can be controlled by spray with Malathion or sevin (0.2 per cent).
2. **Fuller rose beetle** (*Pantomorus cervinus*): This grayish brown weevil attacks lilies and a wide variety of other plants. It is a night feeder, eats ragging areas from the margins of the leaves. It can be controlled by the application of Sevin @ 0.2 per cent.
3. **Stalk borer** (*Papaipema nehris*): Stalk borer attacks asters, hollyhock, phlox and lily. The larvae are about 2.5 cm long. To save valuable plants it may be worthwhile to slit the stalks open and destroy the larvae feeding in the pith. It can be controlled by spray of methoxychlor (0.1 per cent).
4. **Bulb mite** (*Rhizoglyphys echinopus*): Marked injury to lily bulb results from infestation by mites, which is made easy by the loose structure of the bulbs. The mites feed around the basal plate of the bulb and destroy the roots. Bulb should be stored at 1.6°C, since this prevents feeding by the mites during storage. Soaking mite infested bulbs for 24 hours in a solution containing 1 g of 25 per cent wettable tedion in water will also provide control.

Diseases

1. **Foot rot** (*Phytophthora cactorum*): Infected plants turn violet brown at the base of the stem and then the infection move upwards. The leaves become yellow acropetally. The severely infested plants fall over due to the collapse of the stem. Drainage is very important for preventing this disease. The affected plants should be removed from the sight and destroyed. It can be controlled by the treatment of bulb for 2 hours at 39°C in hot water followed by a 30 minute dip in 0.2 per cent benomyl eliminates infection and stimulates plant growth.

2. **Fusarium scale rot** (*Fusarium oxysporum):* The initial stage symptoms are foliage yellowing and wilting. Removal of infected plants and soil treatment with steam sterilization has shown good disease control. Hot water treatment of bulbs for 2 hours at 39-40°C followed by a 30 minute dip in benomyl eliminates infection and simulates the plant growth.
3. **Botrytis blight** (*Botrytis cineria*)- Initially appears as small circular or elongated, sunken spots somewhat water soaked. Flowers buds are shriveled, distorted or disfgured depending on the severity. Use of disease free stock and resistant plant material are recommended for healthy flower production. It can be control by mancozeb (0.2 per cent).
4. **Soft rot** (*Bacillus lilii* and *Pectobacierium carotovorum*)-The characteristic symptoms are soft and wet decay of bulbs. All diseased bulbs should be discarded. Healthy bulbs should be planted in a well drained soil. Care should be taken for avoiding injury to the bulbs while lifting from soil.
5. **Tulip breaking virus**: This virus produces severe leaf mottling distortion of the flowers sometimes giving rise to notched petals and a general reduction in growth. Transmitted by the aphid *Myzus persicae, Macrosiphum euphorbiae* and *Aphis fabae* diseased Darwin tulips commonly are the source of infection. It can be control by Spray of rogor at 1.5 ml/litre of water at regular interval.

Physiological Disorders

1. **Bud blasting and abscission**: Flower bud blasting is characterized by weathering and bleaching of the flower bud, followed by necrosis and bud drop. This may occur at any stage of bud development. Abscission usually occurs when the bud is about 1.2 to 2.5 cm in length. It is associated with low light intensity and short photo periods. Bud abortion also occurs in the late spring and the early summer due to high temperature. Abscission can be eliminated by providing artificial light about 450 watts m one month before flowering.
2. **Leaf scorch**: It is also known as leaf burn and tip burn. A period of bright sunlight after prolonged dull weather can cause sun burn or sun scorch on the leaves of some cultivars. Scorch is noticed particularly at the critical visible bud stage and will produce white bands across the leaves that eventually become necrotic. This disorder may also be induced by low calcium content in the young expanding foliage. A foliar application of 1 per cent calcium chloride just before the visibility of buds has been found successful in reducing leaf scorch. Leaf scorch may also be induced by high fluoride level in the soil, water or air.

Index

www.ingramcontent.com/pod-product-compliance
Ingram Content Group UK Ltd.
Pitfield, Milton Keynes, MK11 3LW, UK
UKHW021010290726
14059UKWH00001BA/65